SEASONAL

ED
ENGLE

SEASONAL

A LIFE OUTSIDE

PRUETT PUBLISHING COMPANY
BOULDER, COLORADO

First Edition
1 2 3 4 5 6 7 8 9

R00715 85269

Library of Congress Cataloging-in-Publication Data

Engle, Ed, 1950–
    Seasonal / Ed Engle.
        p. cm.
    ISBN 0-87108-780-4
    1. Engle, Ed, 1950– . 2. Foresters—United States—
Biography. 3. United States. Forest Service—Officials and
employees—Biography. I. Title.
SD129.E5A3   1989
634.9′092—dc20
    [B]   89-35042
                                                              CIP

This book is dedicated to
my mother, Bernice Engle, and
the memory of my father,
Edison G. Engle, Jr.

Contents

PREFACE

Stories are often like the seasons, and, as a writer I sometimes feel like a weatherman. I start by poking my head out the door and looking up at the sky. To a certain degree things are predictable, but it is the quirks and odd creases that make it interesting.

Seasonal is a collection of personal narratives about the twelve years I spent as a seasonal worker with the U.S. Forest Service. The narratives took place in a variety of places—northern Idaho, Colorado, California. I moved around a lot, like most seasonals. In fact, I think moving around may have as much to do with these stories as anything.

I thought I knew a lot more about the outdoors than I really did when I started with the Forest Service. I mean, heck, I had a Kelty backpack, some pretty hard backcountry mileage on my boots, and a degree in biology. It turns out that I went into the woods twelve years ago with a lot more answers than I came out with.

The few things I've learned are simple—as a group foresters tend to be taller and skinnier than most. Their knees go bad early on. On the other hand, loggers tend to be short and wiry. There are times when a good pair of boots means more than good sex. And that no animal or plant in the woods ever compromises just because things are hopeless. This is what separates them from bureaucrats.

A number of the narratives in this collection literally took on a life of their own. It might be that I started them off one way only to find the ideas tugging at the reins to go off in another direction. In that way they were something like any decent walk in the woods. I like mystery and high adventure even if it occurs in a story that I'm supposed to be shepherding along.

As much as the woods themselves, I have come to respect the company of the people who work in them and live around them. I don't mean everybody. I'm talking about the ones who for one reason or another have made a commitment that almost always puts them on the back side of the American Dream. Working outside is often dangerous, and the pay is poor. There isn't much social status to it, either. The point here is that a way of living may be more intoxicating than just earning a living.

It's customary in introductions to acknowledge the men and women who have made everything possible, and I'd like to do that here, but there will be some deletions. When I started work on this project I contacted a number of friends who live in out of the way places. They all thought it was a great idea, but most of them mentioned that I might not want to use their real names in the book. They're probably just shy, the way woodsmen and woodswomen get, but I'm not sure if there aren't a few of them on the wanted list. So thanks to all of you.

Among those who have already served their sentences, I'd like to thank John Gierach for unswerving support throughout the whining and crying stages of the book. Jim Pruett has also provided valuable support during the entire process. This book could not have been written without my wife, Monica's, support and editing skills.

Ed Engle
Lyons, Colorado
1989

SEASONAL

Four Mile Run formed the
northern boundary of the neighborhood where I grew up.
We never called it that but referred to it simply as "the
creek." The creek, more exactly, meant not only that ribbon
of water but the strip of neglected land around it where
the Old Dominion Railroad and later the Chesapeake and
Ohio right-of-way came through. The train wasn't much,
nothing more than a few switching engines, but the right-
of-way proved to be a fine tangle of brush that was home
to rabbits, snakes, box turtles, and praying mantises. The
creek provided frogs, tadpoles, and shiners, an occasional
sucker, and the rumor of trout.

To the west along the tracks toward Falls Church and
about a half mile from my house there was a break in the
continuity of the creek. A small waterfall, no more than a
couple of feet high, marked the fall line. It was the place
where the rockier soils of the hill country or piedmont met

the softer soil of the tidewater plain of Virginia. For no reason that I can understand now, I had the notion of the piedmont being a better place to live and even went to the extreme of imagining that I had sighted the Blue Ridge Mountains once when I was near the waterfall.

The suburban view from where we lived in Arlington, Virginia, across the Potomac river from Washington, D.C., was crowded, and the only relief other than the creek was a parcel of acres off Wilson Boulevard. There was enough hardwood canopy and brush on them to support at least one fox, which made itself known to me when I was sneaking around in there. We formed a sort of brotherhood in the covert world of the woods because there was no way that we dared be found out in that place. In the center of the woods stood a big, white frame house with a huge red, white, and black swastikaed flag unfurled over the front. This was where George Lincoln Rockwell lived and it was the headquarters for the American Nazi party. The signs all over the property vowed death to any intruders and the land was occasionally patrolled by a tribe of miscreant storm troopers.

Once on their rounds they flushed me up, but I knew how to run in the woods by then and had the added advantage of knowing the trails through the brush. I outran them and watched the blur of the swastikas banded on their pumping arms from the safety of neutral territory outside the perimeter. I kept up with that land into high school and for a while maintained a stash of gin, buried in a wooden box back there, that I dug up every Friday night. The Nazi danger had passed when Rockwell was gunned down in front of the barber shop up the street from my house. In the early 1960s being chased by Nazis was as close as I came to adventure in Arlington, Virginia.

Wilson Boulevard did have other interests. Up from the Nazis was the mansion where Jane Meadows lived for a while. I remember seeing her in the supermarket, when she was still flushed by her fame in *The Honeymooners*, and hiding behind dark glasses. Her hair was really red which, of course, I could have never imagined since no one had color television. The swimming pool behind the mansion was a good place to try and ease into very late at night when the humidity hung over the suburbs and we couldn't sleep.

The most interesting person on Wilson Boulevard lived at the end of a string of houses each of which occupied what seemed like a good measure of land. These were the older places that had been there before the subdivisions were built in the early fifties. Mrs. Inscoe lived in the last house, which was next to Powhatan Springs. Powhatan Springs had at one time been known to provide the best water in the area. The rumor was that in the late 1800s it was used as the source of drinking water for the White House. Over the years the old springhouse had deteriorated and was little more than a rocky home to the kind of salamanders that can only live in very cold water. A community swimming pool had been built on the hill above the old spring.

Mrs. Inscoe was already old when I began delivering her newspaper. We sometimes talked because her husband had just died and she was lonely. I'd sit on the steps of the porch and listen to her stories of when they'd first come out on Wilson Boulevard in 1915. The road was dirt then and each day one of them had gone down to the spring to fetch water. She talked about days when spring rain was important and autumn harvests a time to celebrate. At the time I'd been fooling with the idea that working outside might be more interesting than being inside and gone as far as joining a Boy Scout troop that periodically marched off into the Blue Ridge Mountains to camp. Mrs. Inscoe left me with the thought that the events of a life might best be tracked in the context of the seasons. The government across the river and the subdivisions that had spread over the hills of Arlington were all right with her but if I was *really* going to rely on something it was Mrs. Inscoe's belief I should throw my lot to the seasons. This was radical thinking for the clanking hubbub of Arlington.

My mother believes that the changes were first noticeable about the time I took up playing the guitar. There was no doubt that by then I'd slipped out of the mainstream of society, and she could hear it, too—I had an amplifier. At that stage my life was a scale of bent notes sliding in and out of the relative minor. I liked the blues, and the good ole boys who played hillbilly and drank late on the hot summer nights wouldn't let me sit in.

The real process of it had started much earlier. My dad used to read me to sleep with the poetry of Robert Service:

> There are strange things done in the midnight sun
>> By the men who moil for gold;
> The Arctic trails have their secret tales
>> That would make your blood run cold;
> The Northern Lights have seen queer sights,
>> But the queerest they ever did see. . .
> Was the night on the marge of Lake Labarge
> I cremated Sam McGee.

I fell away to dreams of hard-boiled men hunkered around campfires in the middle of nowhere. It sounded all right. The idea of that kind of vastness had a certain ring to it.

Dad was a spook. That is, he worked for the Central Intelligence Agency. I never saw his office, never followed him through a day's work, and never talked to him about what he did. I knew that he was a scientist and that the climate was somehow his bag. In school when we filled out papers his employment was always listed as the U.S. Government. The teachers asked for more information and I'd tell them that's as much as you can have. They responded knowingly, "Oh, he works there."

Dad's interests were unbounded. He was the one who brought home the recordings of Huddie Ledbetter and Woody Guthrie one day and a fossil of a seashell the next. When he travelled he carried back tales of jumping from a sauna into a hole in the ice up in Finland, the deserts near Tucson, a winter in Churchhill, Manitoba, where the mittens had a ruff of wolf fur to wipe your nose with, and

4

an airplane's engine burning up, and an emergency landing in Iceland. He brought back fishing rods for the entire family from Europe.

Our common language was fishing. Dad came from a line of fishers. His mother and father, aunts and uncles, and grandfathers had all been fishers. We had a photo album of them, each standing on some forgotten river or dock, holding fish of all kinds, shapes, and sizes. For his part, Dad used to come into my room early, at three or four in the morning, and get me up so that we could be on some secret bass lake that he'd heard about before sunup.

It was in those off times, the odd hours of driving into the darkness of the countryside that we talked. We figured the lures we'd rig up with and what part of the lake to fish first, during the best times. Sometimes he talked about when he was growing up and tried to run away from home to be a trapper in Canada or the hidden cave he'd known on the bluffs above the Susquehanna River. I now think that there is a kind of wildness that runs in the genes of our family. A migrant craziness that dimples the surface like a rising trout that only comes out in the after hours and during the off season. Hurtling through the darkness on those mornings I felt the excitement and was glad I knew someone I could talk about it with. It had to do with being on the move. The act of going and seeing as art. Migration.

Sometimes we just headed out to the mountains with the fishing gear. In the summer we might rent a cabin in southern Virginia and fish in the mornings then take off hiking later in the day. Now and then the hikes were tough, but I learned long distance walking on them. The idea was always one of endurance, lasting as long as I needed to, as opposed to strength, which was seldom required for just going into the mountains. If he could, Dad always engineered a hike that would cross the Appalachian Trail. That 2000 miles of trail was a vision to him. The idea of that distance represented a lot of going to him and that in turn formed the foundation of a philosophy.

One summer we took off to Maine. Our family lived by a lake there for a time. My sister and I paddled across the lake each morning to fetch drinking water from a spring on the other side. Our trips terrified my mother, who thought we were too young to be out there alone. Dad, in his typical way, simply said that both of us were the best swimmers he knew. We continued our routine.

The summer in Maine stands out in my mind because of one side trip we took. Dad had a way of finding an old sidekick to visit wherever we went. It might be someone he'd grown up with, an old fishing buddy, or a friend he made at work. This time we headed way out into the boondocks. Dad knew a Russian defector who lived there with his dogs. Maine had suited the old Russian on retirement because it reminded him of home. He was a wonderful man who was living as close to self-sufficiency as anyone I had ever seen up until then. His working years had been spent in the Washington area, where Dad had come to know him.

After a while he sat me down and told me the story of his white tuxedo. It seems that his duties had led him into attendance at a number of functions around the capital. This was understandable because the government would surely have wanted to show off anyone who'd "come over." The functions were formal and he usually attended trussed up in the confines of his white tuxedo. He pointed out his window to a huge garden when he came to the final line of the story. "When I came here I went out into my field and spread manure in that white tuxedo!" He roared in his thickly accented voice.

The Russian had been at the cabin for a number of years and was known throughout the community. In the winter he would ice skate the twenty miles into town down a nearby river that froze over. He cut his own firewood from a woodlot on his property. When we were leaving he gave me a feather from a bald eagle. He told me that one day six of the great birds had flown over his cabin and he'd seen this feather float down as they passed. It was what they called a primary feather, from the wing, and he took it as a sign. When I got home I stored it in a box with my one other treasure, which was a marshall's badge that one of my relatives had worn when he walked the streets of Abilene, Kansas. He had been there in the wild days.

The Russian was the first link in a chain of subtle conversions. The events that turned me, although I had no awareness of them, were like a hard change in season that grows into the tissue of a tree. I had no idea then that a person could step off the main track, that strange things really do happen in the midnight sun, that going could be a way of life. Years later an astrologer said I might have seen it in the jumble of my astrological chart. There was no grounding. I could see my stars as a flock of wandering, migrant birds. But, at the time, all I wanted to be was a solar physicist.

Harry Grimmnitz was my age and lived across the street. He had a different view from other teenagers because he hadn't been well early on and missed the first five or six grades of school. His father had educated him at home. When I came to know him he was okay and plugging away with the rest of us in public school. We engaged ourselves in what most of the other boys our age were doing. It amounted to skipping school, wrecking cars, drinking, and trying to get laid. The exception was the summer that Harry went off to New Hampshire to work in a country hotel. He came back with a blocky pair of European mountaineering boots, a backpack, and a copy of Colin Fletcher's *The Complete Walker*.

We began hiking and backpacking whenever we got the chance. Sometimes we headed to the Great Falls of the Potomac River and scrambled on the rocks that were slimy from the spray of the cascading river. We tried to hike the river from Georgetown up to the falls but were turned back by thickets of briars and honeysuckle that demanded a machete. Our best times were the trips west, to the Appalachians.

Once, we got way off the beaten track in West Virginia. After a hot, humid morning of climbing up ridges and down into valleys we settled on a campsite on a grassy knoll. Our setup was straight out of *The Complete Walker*, which had become our bible. A line of thunderstorms cooled the evening and we went to sleep under a remarkable canopy of stars. Harry swears a bear visited our camp that night but I slept through the ruckus. Being chased through the woods by Nazi storm troopers years before had substantially lowered my fear of any animal, other than man, that roamed the forest.

When we came out of the hills we found that Harry's car wouldn't work. We walked to the nearest cabin and found an old guy to help us. Although he'd never seen a foreign make like Harry's car he lifted the hood and had it running in a few minutes. I remember him mentioning, as we left, that the reason America's cars were so good was because of the competition. He reeled off the names of Nash, Packard, DeSoto, Whippet, and a number of others I didn't even recognize as examples of the fierce competition that in his mind formed a basis for a technological kind of natural selection that made for great cars. We didn't tell him that none of these companies had existed for the better part of twenty years. It occurred to us that he might not have been to town since the days those extinct models had plied the roads like a ghostly herd of bison.

One of our last hikes together was down a trail near Corbin Hollow in Shenandoah National Park. We followed a brook that rounded a bend and dropped into a large, deep pool. It turned out that the bend was more than a twist in the topography. The pool was full of naked, laughing hippies. These longhairs were the vanguard of a wave of thought that was boiling up out of the West. There was an idea in the air that people could subsist off the land, be one with nature, and return to something that wasn't found in the woodlots of suburbia. Our parents, who almost to a person had migrated to the cities from the farming country, had talked about it. They'd said they didn't like it. Harry and I put our backpacks aside, stripped, and jumped into the water.

Harry was a year ahead of me in school and I drove
him to college in Pennsylvania that fall. He flunked out
with finality in a couple of semesters and headed out to
California to participate in a walk for ecology. When I saw
him again he was wandering around outside his parents'
house in the garb of a Franciscan monk. He told me strange
tales of days spent hiking with old California hipsters and
younger longhairs. Sometimes all they wore was their
mountain boots. He had a tan all over. A few weeks later
he took off for the country in a banged up VW microbus.
I wouldn't see him again for years.

I ended up in Boulder, Colorado, after some bad years
in the Midwest. My path hadn't been a straight line, but
the fact that I would end up in the West was inevitable. I
had taken to the idea that a person could live off the
country in the Rockies and everyone seemed to be ending
up in Boulder. I decided to finish my degree in environmental biology there. I drove straight to the campus after
a summer spent on the top of Mount Rainier. I'd fallen in
with a pack of mountain climbers and finagled a deal
where I was part of a team that read thermometers placed
around the cone of that volcano. We lived up there in the
snowfields at 14,000 feet for three months or so. The point
of the project was obscure to me at the time, but years
later when Mount St. Helens blew, I got the picture.

The idea in Boulder during the late sixties and early
seventies was to look like you had just fallen off the turnip
truck, or better yet, just gotten back to town from a month-
long buffalo hunt. It was a university town full of people
who somehow looked indigent. Blue jeans were sewn over
with leather or worn denim. Old jean jackets could be
sold at exorbitant prices, as if they were rare herbs, and
men wore their hair long, tied back in pony tails, unlike
the longhairs from the Midwest who wore it straight.

Things still sometimes got rough between the cowboys
and the longhairs. Having long hair was a good course of
study into the mechanics of prejudice, which most of us had
experienced only as theory. Real life dramas occasionally

played themselves out in the crazier bars and darker alleys. In time, the longhairs overwhelmed the cowboys and the town itself. Herb shops and health food stores sprang up, a guy started selling a tea called Celestial Seasonings on street corners, the Hi-Lo Bar, a notorious cowboy hangout, went disco, and rumors spread about cheap land, home-steading, and tipis.

Boulder not only represented a growing environmental concern throughout the country but a ruffle in the suburban consciousness of America. There was doubt that all was well in the middle class. We clamored for a simpler life— houses with wells, outhouses, acreage, and gardens. Getting into the country was the answer. The hyper-charged atmosphere of Boulder was an exciting center of commerce for thousands of transients hitchhiking around the West, their backpacks crammed full with books on Zen Buddhism, poetry, and philosophy. I liked what I saw and let my hair grow.

The biology department at the University of Colorado had trouble keeping up with the flood of students wanting to study ecology. The department, up until then, had been more a backwater pushed to the side by the rush to the hard sciences in the sixties. The professors, who were good at what they did, worked quietly and thoughtfully at the kind of projects that interest university biologists. The horde of new students discomfitted them. There wasn't enough space in the classrooms, so they rented the Fox movie theater for the introductory biology lectures. Upwards of 2000 students attended the lectures where the professors sweated under stage lights.

I went through the motions at the university. The idea that a multitude of species and the environment could be neatly packaged into ecosystems, biomes, ecotones, and even ecotypes was seductive. The natural world seemed to be laid out like a huge football field with ecotonic yardage lines separating biotic communities. There were energy equations that could be calculated to show a kind of equi-librium even in the most confusing situations. Populations followed curves that jumped up and down according to resources available. They were the kind of ordered explanations that made sense to a class of students with a basically suburban mindset. On a technical, more adult level they also made the utilization of natural resources by industry a little less complicated, but we hadn't figured that out yet.

Outside the classroom, which didn't take much of my time, I was running with a seemingly odd mixture of rock climbers and poets. Bill Putnam had gotten me interested in technical climbing when I was in Michigan. He was a superlative climber who had taken up the sport when he was a cadet at the Air Force Academy in Colorado Springs. He told me that he'd become deeply disillusioned with the Academy in his second or third year and gone to the Academy shrink. Bill told him that he couldn't stand it anymore and was going nuts. He needed out. The shrink told him that every cadet gets crazy now and then but if he stuck it out maybe he'd get jets. Flying was the draw. Fate intervened when Bill got mangled badly playing lacrosse and was discharged. He went to Michigan to work on a degree in psychology.

We used to go out to a forty-foot cliff when we lived in Michigan called Grand Ledge. As a climber Bill was in a separate class but he showed me the basics. When he decided to move from Michigan to Boulder, where there was world class climbing in the canyon west of Eldorado Springs, I followed him.

I never climbed with Bill in Colorado. He was busy putting difficult new routes up that were beyond me. I found a partner and settled for classic climbs with names like the Bastille Crack, the Ruper, Rosy Crucifixion, and the Maiden. Once we struggled up Gorilla's Delight. In the climbing guides this was classified as a 5.9 climb, which at the time was considered fairly difficult. When I told Bill he was delighted. He celebrated the fact that I was finally getting into things that held interest. I had been scared shitless the whole climb. It was the only 5.9 I ever did.

SEASONAL

The poets were loosely organized around a bear of a man named Jack Collom. We held readings and spent time in workshops scattered between our houses where we read and critiqued each other's work. One winter Jack held a workshop in the mountain town of Black Hawk. He worked us hard, driving constantly for precision. He didn't want to hear about a poet's personal problems or deal with slop. Jack just wanted to hear good poetry.

Besides being a masterful teacher of the language, Jack was the best biologist I knew. He didn't have much formal training in the *science* but relied on an intuitive understanding of the natural world, which he gained through observation. His specialty was birds. He knew the hawks and warblers and where to find them when the wind blew hard or the rain was coming in sheets. The skills were hard learned because he was the one scuttling around out there figuring out what wasn't printed in the guide books. He had a way of putting the lives of birds into a context of time and space that was a step beyond evolution and the classes of ecosystems.

Jack showed me that a man can go absolutely, full bore, all out. It might not always be pretty or even instructive, but it's an idea I've held on to. It gets right to the heart of writing good poetry. There must be precision in the face of fear, which also explains why poets might run with rock climbers. Jack or Bill would call it a kind of confrontation with odds that aren't necessarily appealing.

These are the threads that can be pulled together into the kind of fabric that sets one man's life apart from another. I'd slipped from the mainstream and decided on a life outside. I wanted to feel the seasons come and go. Be out where it was happening. It had to be better than television.

I graduated from the university and took a minimum wage job as a landscaper around Boulder. I laid out that first winter. The next year I worked a quarry near Lyons, Colorado, then moved on to a horse ranch. One winter I ended up in Mexico. I tried making it in Eagle's Nest,

New Mexico, and went broke. Finally, I got a seasonal job with the U.S. Forest Service. They needed someone to man the Visitor's Center on Monarch Pass near Salida, Colorado, and do some trail work on the side. It got me off the prairie and into the high country. I was walking country few people ever got to see and doing it every day. It was grunt labor for sure, but I had a suspicion that if I just stayed outside long enough something would come into me.

There are no singular reasons for a person ending up under a bridge, out in the woods, or in an executive suite. The times, like the weather, play hard into it. There was an open, robust, and unbureaucratic environmental movement then that could have pushed me outside, but it certainly wouldn't have moved me toward the Forest Service. There was a bad war going on that could have been held as grounds for an escape to simpler ways. But it could just as easily have been as simple as a Robert Service poem, Huddie Ledbetter's blues, or a crazy gene that turns a man migrant as much for the act of going as for the act of seeing. There may be no better reason than a vee of geese heading south or a break in the creek known as the fall line.

TIMBER BEAST

I don't like the phrase "New Age" very much, but it might fit the store that I went to right after work one day. It was on the main street in Durango, Colorado, and was stocked with things like Peruvian sweaters, a few shirts from India, good paintings of the Southwest (maybe a Doug West repro), and a lot of books about being outside both in the 1800s and now. There were adventure guides, trekking guides, and birding guides. I like these kinds of stores because they remind me of when I was younger and there was a wildness in the air. Maybe they're closer to "alternative stores" or "whole earth" stores; anyway, I'm usually comfortable in them.

It was summer, so my work clothes consisted of a white long sleeve cotton shirt from the Methodist Thrift Shop, blue jeans, and logging boots. The shirt, which kept the sun off me and was also cool with the sleeves down,

15

was heavily speckled with dots and splotches of red, blue, and yellow paint.

"You must be a painter," the man behind the counter said.

"Sort of," I replied.

"What do you do? Landscapes?" he asked. I guess I looked like the outdoor type.

"No, actually I paint trees so the loggers will know which ones to cut down," I said.

He eyed me with the kind of honest disgust most often reserved for people like Richard Nixon and James Watt. I expected him to grab a copy of one of Edward Abbey's books, which were on a nearby shelf, and hold it toward me like a silver cross in the face of a vampire. I would be the enemy. A "tree pig," as Abbey calls them. In the Forest Service we were called Timber Beasts—the seasonal employees who went out every day, rain or shine, to meet America's growing timber needs. The funny thing is I *like* Ed Abbey—he grew up near my neck of the woods.

When I hired on for my first season with the Forest Service in Salida, Colorado, the idea was that I'd man the Visitor's Center off Highway 50 on Monarch Pass. They figured my degree in biology might be helpful in explaining things to the tourists going over the Continental Divide. Mostly I sold maps and counted visitors on a little clicker. I read the *Rocky Mountain News* cover to cover and watched the clouds scuttle by the trailer there at 11,312 feet. Now and then one of the friendlier tourists offered me a drink, which I always took them up on, out behind the trailer.

As the season progressed my hours at the Visitor's Center were cut and I began spending more time out on the trail crew and occasionally with the timber crew piling slash from old timber sales that would be burned when the snow flew. I didn't know exactly what it meant to be a seasonal that first year, but I came to learn that in terms of work I was limited to 180 days a year. This meant roughly nine months work at 20 days a month if they kept

16

me on up to the limit, which wasn't likely. Most seasonals lasted five or six months depending on when the weather drove them out of the mountains or when they needed to return to college. Seasonals weren't eligible for any of the retirement or health plans that permanent employees got, but they did pay into Social Security and accumulate sick leave and annual leave days during the time they worked.

The pay varied according to your grade, but it was actually average or above average in comparison to the standard wage paid in most of the small and sometimes remote rural towns where many district ranger stations were located. In Colorado, most of these towns depend heavily on tourism and a mishmash of agriculture, mining, and logging to get by. Work was seasonal and paid grunt labor wages unless you could get in with some of the local good ole boys.

Toward the end of my first season I had a talk with Milt Robertson, who was the district forester. Forester is a kind of generic term that nowadays can refer to specialists in recreation, range, wilderness, or timber. What it means is that the person received a four-year degree in forestry and is classified in that series by the Office of Personnel Management. Salida was a small district and Milt wore two hats—he took care of both timber and range for the district. Bigger districts that cut more timber may have three or four foresters just in the timber shop. The Forest Service tradition is in timber, and most districts will have at least one forester whose job it is to manage the district's timber cutting and tree planting operations.

Back when I talked with Milt, timber was still king. He sat me down and said that if I was interested in working a longer season I'd try to steer my way into the timber program. It just so happened that he needed someone that fall to do some timber inventories on the district. I told him I'd do them, and within a week I was up in Bailey, Colorado, at a Stage II timber inventory training session and on my way to becoming a timber beast.

The Stage II school was mostly made up of the seasonals

who would actually go out into the field and survey the timber. Stage II was simply the designation for the most intensive type of survey the Forest Service did on its timber. Stage I surveys were less hectic and designed more to look at long-term trends in timber growth. The idea behind the survey was to randomly sample individual stands of timber. This was done by laying a grid over an aerial photograph of the forest and marking "points" or plots at predetermined distances from each other. On the ground you used a compass and accurate foot pacing to get to the points.

Once a point was reached and the location double checked against the air photo, a number of measurements were made on the trees there. Heights, diameters at breast height (DBH), age, diseases present, damage, defects in the wood, the fullness of the crown, and relation of the tree to those surrounding it were all taken. Seedlings and saplings were also measured and recorded. Habitat types, determined by noting the flora under the forest canopy, were recorded.

I liked inventorying timber better than any work I had ever done and probably more than any I will ever come to do. Sometimes we worked in pairs, but most often it was a solitary occupation. Each morning I suited up in a cruiser's vest that was loaded with clipboards, air photos, maps, timber defect and damage charts, my lunch, and plenty of pencils. I strapped on a belt with various measuring tools and an ax around my waist.

In the Rockies it was almost always possible to find a single tree or clump of trees on the low-flown air photo that could be located on the ground. Once found they became a reference point where an azimuth and distance could be taken from the photo to the day's first point, which was pin pricked on the photo. Pacing was critical, and I learned over time to compensate for going uphill and downhill, both of which almost always added paces because the distance measured off the photos was a horizontal distance unaffected by the bumps and grinds of topography. Simply put—the steeper the grade the more paces I added.

The randomness of the point locations in the forest, necessary for an unbiased statistical blowup of the information I gathered, had a beauty of its own. It forced me with straight line accuracy to go places I wouldn't have dreamed of going to for any other reason. We are a goal-oriented species, and although our history is cluttered with straight and curved lines that end at landmarks like fishing holes, mountain tops, oases, and taverns, we seldom set a course, walk a certain distance, and simply stop and look around.

At each point I went through the almost mechanical series of measurements. Like a kind of physician, I looked closely at each tree. I used a long, hollow steel tube with threads on one end and a handle on the other to bore into the trees. When I screwed the bore into what I thought was the center of the tree I extracted a core. By counting the rings of the core I found a tree's age and the years when it had grown well or done poorly. Sometimes I found rot that had come in through a wound or on the wings of a wood boring insect. When that was done I needed to stand back from the tree and look for crooks or sweeps or forks and signs of disease. I wrote it all down on the charts in my clipboard.

Over the years and through thousands of plots I have come to see trees as individuals. It might be that a single giant Douglas fir is preventing a smaller ponderosa pine from receiving enough sunlight to prosper. An aspen might provide the shade necessary for a subalpine fir to get a start. That fir might ultimately grow to cause the death of the aspen. There may be five trees in a clump with just one of them dominant, and that will be the only one that survives. There are sick trees, healthy trees, dominant trees, and subordinate trees. Once I saw a huge Doug fir towering alone in a stand of aspen. All around it were hundreds of fir seedlings. They were that individual tree's progeny, they could have come from nowhere else—I saw them as that tree's children. I hadn't perceived trees as parents or individuals before then, but that fir with the shade and protection it provided by moderating the climate on that little patch of ground was providing in the same way all parents provide.

I have run into spooky ground seeing trees as individuals. It isn't something you talk openly about down at the local bar because they cut your drinks off—let alone if you begin referring to some trees as not only individuals but personal friends. There are some that I check up on whenever I'm in their neck of the woods—just to see how they are doing.

All of this runs amuck of science and its siamese twin, technology. The forest out there is very much like many of the things we know well. It is not the peaceable kingdom. There may be one tree screwing another out of light, or water, or minerals. In our world these transgressions might be called acts of aggression and the cause for war. I'm not sure that war doesn't occur in the forest; it's just all in a different time frame.

The smooth curves and ideas of orderly succession and evolution in the forest make about as much sense as those same curves would make if overlaid on our own human set of circumstances. It isn't that they aren't true, it's just that we don't want to hear about them. A scientific approach works well for populations where statistics can be gathered, but any single individual in that population can remain forever mysterious. What the scientific approach does do is help define and categorize resources like forests to make them more accessible for harvest. This is how entire stands of timber are brought "under management" by the Forest Service.

Like I said, too much time spent in the woods alone, looking at trees, can get spooky.

If I had any illusions about my inventory work they were quickly dispelled. When the data was processed it came back as a thick computer printout. Some of the foresters I have worked with don't even know how to read much of what's there. What they do know how to read is the bottom line, which is volume. This is how many board feet an acre a stand possesses. That in turn tells them if it will be enough to meet their cutting targets.

Target is a nasty word in the Forest Service nowadays. The new word is "goal." Simply put, the target or goal is the number of board feet of timber that a district is expected to put up for sale during a season. We called it the "Big V." Not meeting a timber goal wasn't the end of the world, but it could reflect poorly on a ranger's rating, which in turn would reflect poorly on the forester's rating.

There was a more subtle illusion that went along with the inventory work. It was not hard to believe that my solitary "cruises" through the forest would be an end in themselves. That when the numbers and squiggles I put on the data sheet were finished nothing would change, that the forest would remain as I had seen it. I could imagine that the information I gathered would somehow enter a vault of knowledge and simply remain there to be pulled out by the occasional forester or biologist who wanted an idea of that country but didn't want to go out and look at it for himself. Sometimes it even happened that way, but the point was that the timber inventories were the first link in a chain that was meant to lead to that stand of timber coming under management, which, of course, meant a timber sale. It might be that it would happen five, ten, or even fifteen years down the road, long after I was gone, but nonetheless my observations would play into any future cutting on that ground.

The flip side of inventories in a timber beast's life was the actual marking of timber for sale. The illusions rapidly faded here also because we went out and actually painted the trees that would be taken by the loggers. In most cases we were the very last people to see the forest as it had been for maybe hundreds of years. Sometimes I thought about it, but the fact was it took years for me to learn to visualize what a stand of timber would look like after harvesting.

With few exceptions timber harvesting philosophy has changed little in the past 80 or 100 years. A big difference is that in the old days loggers often went into a stand and just took the biggest and best trees, which were also the most profitable, and left. They called it highgrading. Today the theory is that you follow a plan where something is left for the future.

Stands are harvested in successive stages where a crop can be profitably cut through a number of entries. When an entire cycle of entries is completed, that stand of timber is supposedly "under management" and can be maintained as a sustained source of timber almost indefinitely. A good cutting plan tries to speed up the natural succession of the forest by avoiding certain successional stages, like grasses and sedges taking over after a harvest, in favor of the introduction of desirable seedlings that a forester wishes to promote either by artificial planting or natural regeneration from surrounding trees that have been left standing.

All plans are long term and can be short circuited in any number of ways. One forester may design a sale with a cycle of 100 years, but when he is transferred his successor might be in a bind for volume or want to make an impression on his boss, so he'll go into the stand and start a new plan. Pretty soon the integrity of the design can be compromised and the stand ends up just getting picked over with no real ultimate plan in mind.

The saving grace in many foresters' minds is their deeply held, unspokekn belief in time. A stand can be totally butchered, but someday, sometime—it may be 500 or 1000 years—that stand will come back no matter how it was treated. In the meantime they can switch to virgin country while they're waiting out their mistakes. This doesn't mean anyone consciously *tries* to screw up. Most of the foresters I've known don't. They work diligently at trying to make things work the way they should. It's just a nice backup to believe time heals all.

I don't know if I buy it or not. I've seen some high elevation, massive clearcuts that make me wonder if things weren't just altered too much. I know that in most areas the timber will come back because I've seen the signs. The scary spots creep up on me slowly, and they are different from the places that some environmentalists point at and shriek about. What spooks me doesn't come from a social or quality of life standpoint, it doesn't even come from the wonderfully seductive frontier wilderness kind of romanticism that we have preserved in the Rockies and I subscribe to, it comes from a more deeply seated intuition, from being out there so much and walking over ground that just does not feel right.

Our job on the ground, marking timber, relieved us of any real consternation over timber theory. Marking timber is production oriented, almost like working in a wonderful outdoor factory. The idea was to get as many acres as possible, marked correctly, as quickly as possible. The forester would give a "prescription" that told us his plan for the timber. It might be he wanted a certain percent of a certain sized timber removed, or maybe a thinning in the understory of the forest. Sometimes sick and damaged trees were "sanitized," often large old growth trees called "punkins" were scheduled to come out. A sale with "punkins," which are more valuable to the timber industry, was easier to market and in the process a forester could include less profitable tasks like sanitation in the contract. Once we understood the prescription we were on our own.

The days I've spent marking all blend into one another. They start early and always with a crew. With the large areas to be covered, marking alone would be folly. Three people are absolutely necessary, but five or six are optimal. I've marked with as many as twelve. We'd begin by getting our "guns" ready. These were the paint guns made by Nelson that operate by a series of springs, cylinders, and pumps activated by a trigger. The gun screws into a can of paint. When everyone was ready a lead man took off to be followed by another crew member and so on down the line. The idea was to take a strip of timber 60 to 100 feet wide and cover every tree in it, painting the ones that met the criteria in the prescription. The staggered start of the crew prevented "holes" where no trees were marked. Each crewman, in succession, tied in with the paint marks of the person ahead. In this way no area was missed.

Trees were painted with a slash at eye height on the uphill and downhill side of the tree so the loggers could identify the tree to be cut. A "stump mark" was made at the base of the tree, downhill side, so that after the tree was cut a timber sale administrator could come through, look at the stumps, see paint, and know the loggers had taken a tree they were meant to cut and not simply slicked off what they wanted.

One crew member would tally trees as they were painted and called out by the markers. At certain intervals a tree was "cruised." This meant that a board foot volume was measured on it and any defects in the wood noted. The cruise trees were later blown up statistically to a volume for the entire sale. This volume was used to establish a sale price.

Some sales went into the thousands of acres and took months to paint. The routine never changed. We drifted through the forest, heads tilted upwards, looking at every tree in our path. There were always judgments involved and enough leeway to save the occasional "punkin" or old growth tree that might be deemed too fine for destruction. We never told each other about out individual deviations from the prescription, but they might be considered a small allowance to a higher environmental consciousness. A direct effect each of us could bring to bear. We took our acts of sedition seriously.

Marking timber presents the forest in an unusual way. We covered the ground as an area so that when we finished a sale each of us had an intimate knowledge of those acres. We knew where the springs were and the almost imperceptible draws that led to them. Over that land there might be an outcrop of rock or a swell that the elk bedded down on. The real secrets of any landscape can only be discovered in this way. We formed a more primitive understanding of "our" acres. It is the understanding that comes to people who really use the land—cowboys gathering stock, hunters, farmers, and sheepmen. This doesn't come easily to visitors in the forest. The winding trace of a trail that comes to a magnificent overlook tells a story of the land that is seen; in fact, it almost paints a picture, but this does not compare to what comes up through your feet when those acres are travelled one step at a time.

We occasionally ran into loggers. Many times they were working a nearby timber sale, and we heard the whine of the chainsaws and the throaty roar of skidders dragging logs from the forest to the decks where they'd be loaded and hauled to the mill. When we met it was almost always as friendly adversaries. The Forest Service was the outfit that told the loggers what they couldn't do out in the woods, and this sometimes pissed them off. As timber markers we never encountered their full wrath, which was reserved for foresters and timber sale administrators, because they saw us as grunts with fairly cushy jobs. This didn't keep them from kidding us about all the trash we marked and mentioning that we might think about nailing a few more punkins for them.

Below the surface, almost to a person, we admired them. Logging, like ranching, was a way of life passed from father to son. It didn't matter that it was going the way of the horse and buggy in many parts of the country. Their lives were ones of quick decisions, danger, and incredible physical endurance. The work, when there was work, was brutally hard.

I worked in the big timber country of northern Idaho one season. The loggers were king there where whole towns depended on timber to drive their economies. It was all new to me, coming up from Colorado where timbering is not so pervasive, mainly because the trees simply don't grow so big in the high, dry climate. One of the first things I remember hearing up in that country was that I would notice most of the women were churchgoers. It was simple enough, because their husbands were loggers, and that was the kind of work where they never knew if their man would come home alive at night.

My neighbor in Troy, Idaho, was a logger. His wife was a beautiful woman with a dream of getting on in the world. She thought about college and maybe a job down in Lewiston after her family was grown. I was there when her son came of age and declared he would be a logger. I helped her wrap up the thick wool Malone pants, the ones that all the loggers wore, that she was going to give him. He was going to work the log decks down the river and she knew that it was as dangerous as any work in logging. It was where a man started, when he was young and quick and still had the reflexes to jump from danger. She had a hard edge in her eyes that somehow verged on sadness. She'd wanted to see him through college, but the "life" had won out. She packed his lunch in a tin pail the next morning and told him to be careful.

If we were the friendly adversaries in Colorado, we were the enemy in Idaho. One afternoon a couple of us decided to go for a beer in Pierce, Idaho. When we got to the bar they knew we were Forest Service and the bartender came up and said, "The bar's closed." There were twenty half-looped loggers sitting quietly in the place. We didn't go in.

That didn't keep many of the seasonals and foresters alike from falling for the romance of logging. They even dressed in the hickory shirts, suspenders, and shortened to boot top length, or stagged, pants that the loggers wore. I could see the draw. The loggers might end up poor, maybe broken physically and living in a single room cabin in retirement, but they were their own men and did as they pleased and paid the price. A Forest Service career man might retire to a secure life, but his job was a never ending series of compromises in the best tradition of government work.

There's a bird called the gray jay or Canada jay that inhabits most of the logging country of the West. The big gray bird is most commonly known as a camp robber because of its habit of seeking out people and picking up items, mostly food, from their camp, or even picnic sites. It seems like the bird's fearless affinity for humans is almost genetic. I have come across them deep in the woods and they fly right up and take food from my hand, on the wing. Somehow they have come to know there will be something for them. Loggers treat the gray jays with respect to the point of gentleness. There is the legend among them that the camp robbers are reincarnated loggers. It's a club you don't get into if you work for the Forest Service.

I worked an aspen sale on Groundhog Point in southern Colorado once. Aspen is just coming into its own in that area mainly because the timber of greater value that is accessible has all been taken. The species has been known as a quick-growing hardwood whose fibers just about cross the line into the softwoods. A soft-hardwood, so to speak. It is the tree that first appears after a forest fire, invading the newly evacuated land by means of suckers left under the ground from adjacent trees. Many of the golden aspen groves that autumn tourists to Colorado ooh and ahh over are in fact the sites of old forest fires. Some of the trees are clones, meaning that they are all exactly the same genetically, having come from the suckers of a previous clone. The trees do reproduce sexually on occasion, but it appears that, at least from a survival standpoint, the cloning is more effective and less messy.

Foresters like managing aspen because the best way to do it is to simulate the kind of catastrophe that seems to get the trees' juices flowing. Since fire isn't practical if the timber is to be harvested, the next best thing is little clearcuts in the two- to five-acre range. These clearcuts need to be separated from each other by a big enough block of timber to prevent windthrow in the remaining trees. The process seems to work, and after the clearcut, the little aspen suckers will be growing in a year or two, forming a thicket of hundreds of trees to the acre. Our job at Groundhog Point was easy. It simply entailed flagging in the small clearcuts, then surveying the boundaries to get an accurate acreage for the clearcut. The timber was then cruised to come up with a volume for the sale.

What struck me on Groundhog Point was the incredible beauty of that stand of aspen. Most aspen tends to be gnarly and heavily infected with countless fungi and rots. The stands deteriorate rapidly and give way to spruce or fir, barring any disasters that open the country up. This stand was clean with very little rot and the trees were big, some towering over 100 feet. The sight of acre after acre of the white-barked, straight-trunked, clean aspen was soothing and wonderful.

There is a great commerce of life under the canopy of an aspen forest. The ground is thick with grasses and herbs, some of which blossom into splendid flowers like the columbine, Indian paintbrush, and violets. The elk used the benchy country of Groundhog Point as a calving ground and their beds matted the grasses down neatly. Now and then larger than average mule deer travelled the worn game trails.

We took this all in as we did our work and could not be anything but happy. We finished just as the snow blew in that fall and left for town. Most stands take about a year or so after marking before they go up for sale and logging starts. The match company in Mancos bought the Groundhog Point timber to be made into the kind of matches that I call "barnburners," or kitchen matches.

I didn't get back to Groundhog Point for three years after the logging began. I guess I didn't really want to see it, but they kept raving at the office about how well the reproduction was coming in on the clearcuts, so I decided to swing by one afternoon. They were right. In the oldest cuts there were hundreds of aspen already three feet tall. Except for the fact that they were shorter than the trees around them, it was hard to tell there had ever been any heavy equipment in that country. The sale was doing what it was designed to do. Timber was supplied to industry, browse was opened to wildlife in the new clearcuts, and the seedlings were coming in quickly and growing thick. It was the kind of sale that foresters slap each other on the back about. It made me sick.

I knew that it would all grow back, and actually from the viewpoint of the aspen, as a species, it was a sweet deal. Their time on Groundhog Point had been extended. The ongoing successional march toward spruce and fir was delayed. I missed the continuity of the clean white trunks fanning over the flats and the lushness that grew under them. The country was trashed with muddy logging roads and the slashy debris from the sale. I'm no good in geologic time frames; if I were I could see that the earth will probably go on, with or without humans and even with or without aspen. What bothered me on Groundhog Point was that I had been part of such a drastic change, one that I will not live out, that I had played a role as powerful as the wind or fire or plagues of disease. This is heady work and it is what we do, but I have wondered if our consciousness of it all might not disqualify us from acting as the right hand of God. I am a great fan of the apparent amorality of the forces of nature, but then again I don't know how trees or the wind think.

I got into timber because I wanted to be out there. I wanted to be touched by the events outback that occur in a moment. A tree that falls on a perfectly still afternoon. A junco that I saw slip on the ice. One insane deer. I have come to like things that don't fit neatly under a bell curve. Being out there night and day has led me to distrust the order that we impose on the world. There is a Spanish saying I heard in southern Colorado. It translates to something like, "Take what you want and pay for it." For me I ended up a timber beast. I paid some, but I know what it means when I see someone strike a barnburner into flame, and my heart is not in the business so much anymore.

OLD GROWTH

Salida, Colorado, is the sort of western town I would have chosen to grow up in. The translation of the name, like all mysteries of language, is vague and imprecise. Given a traveller's direction, it could mean entrance or exit. If one English equivalent need be applied, "gateway" might best suit the town, if not the word.

Salida lies in the upper Arkansas River valley at the southernmost end of the Sawatch Range where the river gathers in its south branch. The valley broadens here on a plain of sediment scoured from the Continental Divide, then turns sharply and deeply on its way to the high prairie. The Sawatch ends dramatically with Mt. Antero, Mt. Tabequache, and Mt. Shavano, all over 14,000 feet, west of town. The knife-like edge of the Sangre de Cristos take up south of town following the river.

This would be the gateway to a country that grabs

at the air—coming or going. Scenery aside, I would grow up in Salida for a simpler reason. I value the smoothness of the curve that its inhabitants represent. In a more rural America of the 1920s or 1930s, the idea that a town should have children, teenagers, young adults, the middle-aged, and old all mixed into the grid of its streets and on the spokes of the wheel of ranches that surround it was typical. If the people of Salida were a stand of trees a forester might call it multistoried—a complexity of stems with no recognizable age groups within it. It would be a stand characterized by diversity with a propensity toward wildness.

The suburb outside of Washington, D.C., where I grew up was different. The brick, two-storied houses occupied equally sized plats set off in blocks. The idea in the 1950s was to provide housing for the soldiers back from the war and eager to start families and careers. Our suburb had a particular character. My parents were among the upwardly mobile generation of new military officers and government professionals. The houses were peopled by captains and majors and GS-7s and GS-9s. Our community provided a safe calving ground for a population of government workers in their thirties. Had we been another stand of trees a forester would have seen an even-aged stand of two stories—adults and children. A stand with less diversity but under management.

I recall travelling to Florin, Pennsylvania, now and then to see my grandparents. The adventure was in going far afield to an exotic new place. I kept mental notes on what I saw and heard. My observations included the hitch in my grandfather's gait when he went to check the chickens and the tip of a finger mysteriously gone. My great grandfather rocked in front of a broad window and spit tobacco juice into an old peanut butter jar. He fancied the girls, so it was my sister who helped him when he worked on a jigsaw puzzle. Grandma cooked and canned and was the one with a driver's license. I didn't say much in the same way that I would tread lightly the first time I walked through an old growth forest.

I met Charlie in Salida the first year I came on with the Forest Service. That was 1977. The ranger district there was small and lent itself to a variety of seasonal chores. Some days I'd man the Visitor's Center on the top of Monarch Pass, other times I went out with the timber crew or the trail crew. Now and then I kicked in with the campground maintenance crew. The strangest chore I had was working with Charlie. He was the lead man of what the Forest Service called the Older Americans.

More correctly Charlie was a participant in the Senior Community Service Employment Program (SCSEP). The SCSEP had come about through the evolution of a number of government programs that had taken form from the Older Americans Act of 1965. Most had been designed to help senior citizens who were living at or below the poverty level and through the work programs bring them back into the mainstream of life.

I don't think Charlie gave a hoot about the etiology of government programs, and the only mainstream he cared about was the Arkansas River. Charlie was a short, stout, barrel of an Italian whose folks had worked their way up the Arkansas Valley in the coal mines. My job had been explained to me behind the closed door of the ranger's office. I was supposed to keep Charlie from working *too* hard. The ranger had said his heart wasn't good and he needed an operation. He was getting up in years. I was the biggest man on the district and they figured if anything did happen I would be able to carry Charlie in.

I learned quickly that *nobody* kept Charlie from working hard. He took his team of old geezers out and they went after their campground work with a vengeance. It would have been enough to straighten the wood piles out and paint the outhouses, but Charlie got into his head that boulders needed to be placed as barriers to the people who decided to drive off the road and through the campground.

More than anything it was the tools that excited Charlie. It was the necessity for tow chains and heavy rock bars and pry bars that sang out to him. Each day he had a new strategy for slinging a big rock up behind the pickup and dragging it into place. When we got it close the whole bunch of us pushed and grunted and pried it to perfection. Charlie was always there shouting directions and letting anybody know if he thought they'd slacked off. None of us got away with anything. And Charlie was the boss. I began to notice that no matter how thick the cussing got I could always look to the center of it, find the hardest spot to be, and there was Charlie. He focused incredible strength on the end of a pry bar.

I learned about Charlie's life in bits and pieces. He had driven a beer truck up and down the Arkansas Valley for twenty years. Hustling kegs on and off the truck had made him strong. He'd also built the route up from almost nothing to a good trade. One day somebody got him mad and he quit. Just like that. He'd only had a few more years to retirement when he walked out. I asked him why he hadn't just sucked it in and waited out his time.

"It's no fun if you can't get pissed off," he said. Charlie was like that. So here he was working for minimum wage three days a week grunting rocks. And it did not bother him one bit.

Charlie took me under his wing. It began over onions. I had a few on a sandwich one day and characteristically he blurted out that a man could do a lot better in the onion department that I was doing. Charlie said he wasn't an Italian for nothing and he would teach me about onions. After work we went over to his house. This was the house that he built himself because that would have to be the way Charlie'd go about it. In the basement he had bins of different kinds of onions. There were sweet onions, big onions, yellow onions, small onions, and white onions, all of which Charlie had grown himself. He gave me a handful of the sweet kind and told me that they were what a sandwich needed. I would have to come back when I wanted to cook with onions because that was a whole new ball game.

Charlie ended up pointing me toward all the best fishing holes in the Arkansas River. There was never any charity to it. He had a way of seeing what a man needed and giving him a way to get it. I needed to catch brown trout. I would sneak down to those spots straight after work and most of the time they were what he'd said they'd be. He never came along, but once I saw his truck slow down just a bit as it skimmed along the highway above the river.

A few times when we went after a particularly big boulder or Charlie had spent a little time giving us a deluxe chewing out I would see a translucent cloud of fear pass over his eyes. A mare's tail of understanding that he had known since World War II. He would get faint, then grab for something to lean on. Once or twice I went over quickly to him to act as a prop. He never let me hold him up. When it would pass he looked at me like I was somehow stupid and that fear or pain was no different than an onion or a brown trout.

I heard later, after I'd left Salida, that Charlie had gone up to the Army hospital in Denver and they had sewn a valve from a pig heart into his heart. He was back at it moving rocks. I had come to a notion when I watched my grandparents that a body could be a cage. A place to wait out the years, but I'd never worked by them cultivating their few acres or in the hen house or picking grapes. For Charlie a body was more like a rock bar or a tow chain or a pry bar.

After Salida I migrated around the forests of Colorado for a few years, then boomeranged up to the Clearwater National Forest in Idaho and ran into John Opresik. John was a seasonal same as me except that he was in his seventies. He'd come out to Idaho on a freight train during his hoboing days in the Depression. He'd found work haying and never returned to his home in Wisconsin. John was a short, wiry kind of man and in his younger years had fallen naturally to logging in that big timber country. He was fond of saying that when he got too old and sick for that

35

at about sixty-five he'd come over to work for the Forest Service to relax a little.

John was a legend around Pierce and Weippe. He'd practically fallen off the train into the area and managed to survive where many locals failed. The other old-timers say that he built his house in the spare time between logging jobs, and when that was done took to getting up at two or three in the morning to work his fields with the help of the headlights on his tractor. That was before he headed out to log at five. If he had time when he came in from the woods he might work at digging a well until the summer light gave way at ten in those northern latitudes.

I worked on a timber inventory crew with him. John never had to get in with the Older Americans program because he was too valuable as a timber cruiser. He worked for seasonal pay each year until he reached the maximum he could receive and still collect social security. It worked well because that was about the time bear season began. John had a way with bears.

I learned that when I worked with John I followed him. He was quick as a cat in the tangle of deadfall and brush that lies under the forest canopy of north Idaho. It was good enough just to keep up. John had a divine affinity with that land and the trees that grew on it.

Our job led us on transects to a number of plots each day. Once at the plots we measured each tree precisely and recorded a number of facts about it. John always paced us through the woods to exactly where we needed to be. At times we estimated tree heights. He'd step back a ways, squint up at some huge ancient cedar, crane his neck a bit, and call out a number. The times I checked him with a tape and clinometer he was within a few feet. Sometimes, when I'd estimate a height wrong, I'd catch John pawing like a bear at the dirt. When he'd gotten my eye he'd squint back up at the tree in question, then paw around a little more, but he never said out loud I was wrong.

John O., which is how we separated him from other Johns on the crew, never used a compass or map. He had a sense for the landscape that precluded angles, straight lines, and declinations. Any little bump or hitch underfoot focused a knowledge into him of exactly where he was. It wasn't that he could point a finger to the exact azimuth needed to walk back to the truck, but more a sense of well being in him that declared he was at home right there and could never be lost.

For those of us from a younger more scientific generation who hover over an aerial photograph, then chart a course, John O.'s way was an oblique meander into the meaning of geography. A place where curves and lines have no future in an equation. It's what was left out at the university.

I scrambled out of Idaho after one season in the rain and landed at the Dolores ranger district in southwest Colorado. The station sits squarely in the town of Dolores in the river valley of the same name. The Dolores River, originally the Rio de Nuestra de los Dolores, received its name back when the Spaniards roamed that country. It translates to River of Our Lady of Sorrows. Although the name is often associated with the Dominguez-Escalante expedition in 1776, there is no real evidence to prove the two Spanish padres had reason to be sorry when they crossed it. The name does provide a gloomy companion for the Animas River, which is forty miles to the east and is the shortened version of Rio de Las Animas Perdidas. "The River of the Departed," or maybe more correctly "River of Lost Souls." Again, it is hard to tell how the name came about, but the story goes that a couple of soldiers drowned in it. Other than names, the two valleys are about the prettiest in Colorado.

The work center for the district, which stood separately on the east end of town, was a knot of activity. Here, the green Forest Service trucks and field crews came and went on their chores over the district's 480,000 acres. The work center itself rang out with the sounds of things being hammered, welded, pounded, routed, and cursed at. In one corner of the yard, which is what we all called the place, there was a metal corral that held a horse or two. The dark hay that filled the corners of the small barn just up from the corral was good for a litter or two of feral kittens each season and sometimes a stray hen that might lay eggs on top of one of the bales.

The rest of that fenced-in plot was scattered with an array of the things deemed necessary for running a forest district—fence posts and barbed wire, concrete picnic table legs, signs, a disker and ripper, lumber, paint, well pumps, and the piles of debris that would be cannibalized, pieced back together, and made to work.

At the epicenter of all this there was a gang of Older Americans who stood as various as the tools, parts, and pieces that were scattered about. I looked for Bob or old Bill whenever I was around the shop. Bob was bent over by arthritis but in possession of the steadiest hand of the bunch. He spent his time routing the messages of the Forest Service into wood. These signs showed up all over the district in a variety of forms: Mavreeso Campground—14 miles, Forest Road 191, SAN JUAN NATIONAL FOREST, Prevent Forest Fires. . . .

Bob often wore the white hardhat that had designated him as an engineer on other jobs in other places. He might ask in passing if you'd ever seen the Astrodome in Houston. He'd had a hand in building that first of a kind covered playing field, and its arching girders of steel criss-crossed with clarity in his mind. The image of its very immensity and the thought that men could build a structure so large that clouds and rainstorms developed in its heights replayed constantly in him. That he could be part of it all, by the laying of his hands on steel and wood and glass, was a source of veneration to him.

Bob had no time for the slow grinding of government gears and cogs. His was a world of production, work loads, and schedules. Once or twice a week he'd stalk down to the district office and rave about inefficiencies and time lost forever. He'd tell them that if this haywire outfit that they called the Forest Service was a business they'd have been bankrupt and out pounding the pavement years ago. What if they'd been ramrodding the job down there in Brazil when he'd built airfields to get America over to the war? Why, we'd have never made it, he grumbled to himself. After those journeys into bureaucracy he'd come back to the shop, say they wouldn't listen to good sense, and settle back to his sign making. He had a quota.

The question that I burned to ask Bob was the one that I couldn't. There is a code among drifters that discourages asking a man where he came from or how he came to be where he is. I wouldn't call a Forest Service seasonal a drifter, but it is the kind of job that gathers people on the rebound. This is especially true for those of us that are a little older and may have spent some time under the bridge or next to the railroad tracks.

I would ask Bob how it was that he came to be with the Older Americans. Why was it that an engineer, or maybe chief engineer, hadn't put anything back toward old age? Had some devastation torn through his family and bankrupted him, or was it that he hadn't ever settled out? Was it that he came to like following the booms as much as the work itself? I wonder if maybe his feet didn't get moving under him and he couldn't find a way to stop them. All of this is conjecture, the storm that unasked questions precipitate, but if it was true, I'd have a feel for the weather.

Old Bill ranged mostly between the shop and the crew quarters in town. He was the short, wiry kind of man who is easily associated with survival on the leanest terms. I expected to see him wearing his sweat-stained cowboy hat and the pointy, slick-soled type of cowboy boots that a man wears who knows that the time will come when he needs to get out of the stirrups quickly. A real rider. From what I heard, Bill, or Wild Bill as we called him when he wasn't around, had ranched or farmed all of his life. This would add up to a lot of days in the Texas wind for Bill. He was about eighty when I met him. Just the gnarly virtue of that fact put him on the top of the heap of our admiration. The idea that a man could live through so many seasons and still crave work made him indestructible.

Bill liked the physical kind of work. The other men, sometimes twenty years younger, often called on him for help when it came time to heft a plate of steel to the welder or a box up into the bed of a pickup truck. It was his gift to know exactly how to make his body a tool itself. He had an understanding of how a subtle change in the attitude of the wrist or angle of a hip might provide the physics necessary to lever the weight of some object to his will. It had come by way of thousands of bales of hay bucked into submission.

Bill's other skill, which I took as the reason for his health and longevity, was his ability to nap. I might catch him snoozing in his pickup truck if there was no work at the moment. Other times he might drop off quickly in his chair after lunch. His virtuosity was the ability to sleep while standing. I walked into the shop once and noticed he had a newspaper spread out on the workbench. He was standing there, head down, apparently intent on some reporter's story. He was as motionless as an egret poised over an unsuspecting prey and his eyes were closed! They say a newspaper did it to him every time.

The few times I talked with Bill he never failed to mention how much the three days of work a week that the Forest Service gave him meant. I don't believe it had much to do with money. A hardscrabble character like Bill could always manage to get by. It was the idea of work itself and the hubbub that surrounds it. Bill had worked his entire life from the time as teenager when he drove a wagon between the then separate towns of Dallas and Ft. Worth. The idea of sitting around the house was as alien to him as a broom or rake moving itself.

My notion of Bill's indestructibility was shattered in the late fall. I was working indoors on a mapping project when I heard he'd had a stroke. I don't remember who told me the news, but the message I got was that it was sad but to be expected. The guy was getting up there. I worked into the winter that season and remember looking out the window one day. I saw Bill being helped from a car. He levered himself up on the door frame and then was given a walker. When he adjusted himself into it he was apportioned in a way that made me think it was more the walker that needed him than the other way around.

Within a few months old Bill, Wild Bill, was back at the shop. He'd thrown the walker away and recovered completely. I believe he stayed on a couple of more years. When he did finally leave the Forest Service, it hadn't been his idea. He'd slowed a bit, maybe napped a little more often, but it had not crossed his mind to quit work. Down at the office they must have figured he might get hurt or maybe they had rated him on some obscure government scale of productivity. In those foresters' minds, the idea might have been that a magnificently battered old tree can be past the age of usefulness and to the point of just taking up space. Whatever the point, there alone in the office, they gave him the news.

I sometimes went out with a crew of Older Americans in the spring to open campgrounds. Campground maintenance was one of the jobs that fell within their realm throughout the season. Springtime was the big push because trash cans needed to be put out, pumps bolted into place and primed, water purification systems installed, and any damage that occurred over the winter repaired. Once a campground was on line the work load lessened to include more standard maintenance like trash removal, painting, and taking water samples.

If I could work it, I tried to gang up with Joe and Tieg. Joe might have come out of that same wiry, tough mold that had spit Wild Bill out at the turn of the century. He was short and thin as a rail. What the two of them did have in common ended with their equally lined and leathery exteriors. Where Wild Bill was talkative, Joe was quiet. Where Bill was happy enough to labor around the shop or town, Joe needed to be out and away from the bustle.

Joe had been born in Spain and came to the United States at an early age for the sole purpose of riding herd on livestock. It might be that he was Basque, but however it was that the genes of his family threaded their way through time, they had given him an ability to understand the instincts of sheep and cows. None of us knew much about Joe other than that he had cowboyed all his life, which probably amounted to sixty years of stock drives, high meadow summers, and solitude. He had never married and they said he lived near Cortez where a niece looked in on him now and then.

I liked the feel that Joe had for water. When we had to pull a pump in a campground he'd be right in the thick of it, hauling pipe and sucker rods hand over hand out of the well. As we came closer to the final section that held the pump, I often sensed a kind of excitement in him like he was somehow getting close to the source of all the questions he'd never asked. Water was a treasure to Joe; it and grass had ruled him through all the years, good and bad, of following and tending his herds. When we'd finally have a pump lying there on the ground, he'd go up to it and touch it quickly with a finger or two, the way my mother used to check an iron to see if it was hot, then sniff the water and maybe taste its sweetness.

Joe had a way of looking an entire day over. His eyes ranged out to where the sun was, how the country laid, and the direction to go. There was a quiet urgency to his idea that a man was at his best if he was on his way to somewhere else. If his job was complete in one campground he wasn't much for taking a break and shooting the breeze. Joe would end up leaning against the hood of the truck, alone, waiting to move on to that next place.

One day Joe didn't make it to work. The rumor was that he'd gone on a toot. Someone else heard that his room had been robbed of the few things he owned and it had discouraged him beyond working. He had the hair trigger that many of the kind in his line of work have, I know that. Their lot has much in common with the wind.

Tieg was a bear. He lumbered through work with his big shoulders rolled in toward his chest in the manner that those animals have. The size of his hands always caught my eye. He was big in an absolute way, but there was no threat or meanness in it. He came up each season from Texas and stayed with friends in Dolores. The Forest Service and the Older Americans were his way out, a kind of working vacation. Tieg always put a good day in, but he wasn't inexorably mated to work for the sake of work like many of the others were. I always had the feeling that some of the great strength that resided in his body was in storage for things outside of work. Each day he quietly tended that reserve like an angler who saves the best spots on the river for special times and special friends.

The fact was that more than anything Tieg enjoyed the company of women. The quality that he possessed was a natural one in the sense that women also enjoyed his company. He didn't have any secret lines or special moves. His way had nothing to do with coercion, or the winning and losing of the friendship of women. It had to do with an attractive gentleness that seems to link itself only with big men.

When we worked the campgrounds on the West Dolores River I used to eat lunch with Tieg down by the water. Sometimes he'd watch the current as it eddied into a pool or cut its way under the bank and wonder out loud about the women he had known. The good times, the bad times, and the troubles that may have been unavoidable. He was never specific but rather wandered like the tongues of current in the river. For him it was as simple and yet unfathomable as the idea that the water he was watching would make its way to the ocean. Had Tieg been a forester his forest would have been full of mystery, wild and beautiful.

During the second Reagan administration the budget cutters hit upon the idea that the Older American programs could be trimmed. It seemed a good idea. The program was relatively obscure and why not trim the tail of the silent

43

majority. They were unprepared for the opposition that arose. It was nothing like a tidal wave but more a little here and a little there. And it was rock solid. About all they got was the across the board Gramm-Rudman-Hollings reduction. The simple fact was that the Senior Community Service Employment program was one of the very few federal programs that did exactly what it said it would. The only opposition that I've ever seen was a ranger district clerk who figured she was a little too busy with her computer spreadsheets to waste time listening to an old geezer rant about how much money they were wasting and some obscure drivel about airfields in Brazil.

Over the past few years there has been a stream of discussion about the old growth timber in the forests of the Northwest. The Forest Service's argument is one for managed and productive forests. The big, old growth trees are past a productive age. They aren't putting on much wood anymore and at best only the bark on their trunks is thickening. It is doubtful they throw much seed anymore. The question is one of measuring the value of old growth. I would make a case to leave what's left where it stands. This would be a vote for words that don't translate well into our language, stands characterized by diversity, and lives left to the wind. A propensity toward wildness.

ROUGH CANYON

Autumn comes in stages to the Dolores River valley. In town, by the river, it was still summer. The cottonwood where the kids swung out on a rope over a deep hole, let go, then splashed into the river was still the baked green color that comes from a whole season of sun reverberating off the sandstone. The water was as warm as it would get. The scrub oak that crawled between the juniper and pinion pine on the slopes was always the last to turn, and it would be green for a good month or more, probably into the first snow.

Upstream near Stoner, past the West Dolores, the aspen were turning. When the morning sun came in low to the valley it stuttered across the south side and ignited the color in the trees that had turned first. It was more than just backlighting. The frost on the pastures sparkled and the red sandstone cliffs farther up the road near Taylor Creek glowed like embers. The sun was in my eyes and I was glad that I'd cleaned the windshield.

I'd gotten out of the district office before anyone had made it to work and had already been to the yard and switched over into a company rig. The beauty of it all was that I was alone and had turned the radio off. If they needed me they knew where I'd be. Switching off the two-way isn't considered sporting by some of the more militant in the office, but they would not need me and besides the valley was a dead spot, anyway. A few zigs and a zag around the bend and the sandstone took care of any urgency that might come throbbing through the air.

The only time I had ever received a serious message over the two-way was up in Idaho when my wife had been involved in a minor fender bender. By the time the message had been relayed through the office and hopscotched through several other two-ways and gotten to me I was told my mother had died. That was when I gave up anything urgent in the air other than the weather.

Farther up the river near Larry Fitzwater's house autumn was winning out. Whole sides of the steep ravines that came off Haycamp Mesa were in varying hues of yellow and gold. The occasional stringer of conifers that ran through the color only added contrast and a thicker dimension to the season. Larry's place was one of the houses we always watched for when heading up to work in the valley. For us seasonals it could be a refuge from the weather or a place to pick up a tool we'd forgotten at the yard.

Larry had started off as a seasonal out in California in the 1950s. He'd seen trail work in the days when it was all horse packing and the saws were the big crosscuts, or "misery whips," as they called them. He had eventually gotten on permanent with the Forest Service and thought he had a life before he started talking seriously with his friends. In those days the Forest Service was not the place for an up and coming man just out of the military. The cities were booming and that was where the future would occur. He ended up quitting and came into town to take a job at the bank. He will talk to you about his twelve years in the world of high finance but seldom meets you eye to eye when he does it.

"I was going nuts. I was fat and they were giving me pills for my blood pressure, but I had a Porsche," he said. He hung in as long as he could and finally fell into a scheme on a feed store near Monterey. He took all his retirement and five years and helped build the store up. When he'd had enough he sold out and came to the Dolores Valley, bought a place, and came back on with the Forest Service as a seasonal. He still works on the district.

"I just fell into it, I really didn't have anything going for me at the time. It was luck. I slept in my car for a couple of years and just put the money back into the store. All I ever tell anybody is that it isn't worth doing something you don't like. . . ."

He looks into your eyes when he says it. Larry's good at the things that living in the country requires. He knows about gear ratios on tractors and sprockets on chainsaws. He can pack horses and raise livestock. He also knows where the fishing and hunting is in his neck of the woods. He'll say he's a seasonal because he likes it and that is the only way to be a seasonal.

Larry had told me a few days earlier that he'd heard the elk bugling up near Bear Creek. I was heading up the Hillside road, which would take me there, and was anxious to hear that rugged country filled with sound. They say it is a sound that isn't supposed to be there. It starts with a shrill whistle made for the prairie and grasslands where the majority of elk once mixed with the bison before they were run into the mountains. A deeper roar like the red deer in Europe make would carry more appropriately in the thickness of the woods.

The elk's bugle is a sound that always seems to settle just above everything else. The notes don't complete anything. They wash low down into the mind and finally into functions of the body that are autonomous. The sound can be memorized but the effect cannot. It is the difference between a scent in the bottle and perfume on your lover.

In the 1960s the Hillside road was driven into the higher country from the valley to mine the virgin stands of Engelmann spruce up top. The clearcutting that took place over a span of little more than a decade was awesome. Most of the cuts, some in the hundreds of acres, still remain in need of reforestation. It isn't that attempts haven't been made but that in most cases they haven't been successful. The promising sites have been the smaller units that were cable logged and have come in through the natural seeding from the relatively close-by seedwall.

The country up Bear Creek to the Highline Trail on the ridge and north past the Cape of Good Hope and to Orphan Butte was probably some of the best elk country in the state before the Hillside road opened up the access. It is still pretty good, but now and then the old-timers talk about blasting the damn road shut and going back to packhorses and stillness.

The road literally claws its way along the steep terrain up from the valley floor. It is cut and fill the whole way, switching back and forth around ridges and valleys in an attempt to fulfill the grade requirements in the engineer's handbooks. I stopped at a point that overlooks directly into Bear Creek where the slopes bench their way steeply down into the bottom. This is where it's mostly aspen before the timber turns to mixed conifer and finally into the spruce-fir. I'd seen a few calves and cow elk bound down off the side of the road. It is on these benches that the rut takes place. The majority of a herd was down on the first one off the road. It was still early and the light hadn't come into the side draws yet. I heard a bull shrieking and grunting when I got out of the truck.

Most of us carry along a bugle of some kind during the rut. It might be just six inches of PVC tubing plugged to make a whistle with the right notes. Others are more elaborate, but the point is to get an answer. It is an opportunity to tie into the ritual that always seems to settle me

off the ground, above where I should be. I blew the three-
or four-note ascending scale across my pipe and the bull
grunted back immediately. I waited and heard him crash a
little through the timber and nailed the notes again.

It wasn't that he would pull off from the cows and
charge up the road. This morning, with the color all
around, I was just throwing some sound out in a gamble.
A chance to see what I could only hear. Maybe the biggest
bull of my life would come thrashing out to the edge of
the timber with his neck swollen and huge and belligerent
as hell. This bull didn't come—I heard him turn and bellow
a deep throaty growl like the big ones do and then the
cows broke off through the deadfall and I figured I heard
him leave, too. I got into the truck knowing I'd gotten
more than I bargained for. A chance at witnessing grandeur
is always a gamble.

I had logged out at the office my intention of doing a
third year reforestation survey on the big clearcut at the
end of the Hillside road. This was up near Grindstone Lake
where the flats roll into the flank of the La Platas. If you
want to go farther you catch the Little Bear Trail, which is
damn hard to find, then tie into the Highline via a little
snatch of the Grindstone Trail. It is easier to bushwhack if
you know the country.

Those were the words I'd written on the sheet, the
way I'd spend my day in the cause of production. But these
were complicated times in my life and I had another plan.
I'd missed out on drawing my cow elk license in the big
computer selection and had decided to do a little scouting
for bulls.

I had also decided that the place I needed to scout was
Rough Canyon and that I would do it on company time.
I wouldn't call it an actual mutiny because the rebellion was
not really open. No one would know. Maybe it was a
defection, a leaving without consent or permission, from
an allegiance to the work ethic. They trusted that I would
be working, but work is a broad concept when there are
no office walls. I figured I was going to have to go to a

place like Rough Canyon to get my elk when the hunting season began. By the time the hunters from the earlier seasons had run and shot the herds up, any animal that was big enough to meet the four point or better requirement on my license was going to be down in Rough Canyon.

He would be in a place so steep and so thick that no sane hunter would go in. A place you couldn't get a horse into. A place that the guys in the four-wheel drives would motor up to the edge of and gaze over and say, "Yep, that's where the big ones are," then drive on. A place where a man wished he'd never gone when he got to the bottom and started looking at coming out—even if he wasn't packing meat. It was a place where the stillness would be total and the footprints absent. A place that the bugle of a big bull could fill completely to the rim. I needed to go into Rough Canyon for a lot of reasons. You could call them work.

I parked the truck on the edge of the big clearcut and made it look like I was out somewhere on that rolling sea of stumps and sod. If someone came looking they would not find me but only owe it up to the fact that the clearcut was so big and that I could be anywhere. It would be like trying to find a lifeboat bobbing in twenty-foot seas. They would look out on it all and see that it was good and that I must be busy out there somewhere and climb back into their truck and make town by quitting time.

I had grabbed the air photos in the office before I'd left and already charted a course down into it. The clearcut had been made right to the edge. A place where the contour lines lumbered lazily here and there at 11,000 feet, then plunged one on top of another down to 9800 feet at the point just south and below of Devil's Point, which is part of the Triassic sandstone that had somehow oxidized red and been laid open in Rough Canyon. It even went deeper than that into the ancient cobbles and pebbles from the Precambrian, from the ancestors of the Rockies, maybe from Uncompahgria, the island range when the sea had been inland. I worked my way down toward the side of an avalanche chute hopping along sharp ridges.

I was on the northeast facing side of the canyon, which made the going easier. At first, it was thickly timbered in the spruce-fir that are found on these slopes. There are a few big trees, but the majority always seem to be the saplings or poletimber that timber beasts call "pecker-poles." They are appropriate on a steep slope like this, which is no place for a tree to grow tall enough to be blown over by a crazy canyon wind or snapped off by an avalanche. The key to it is a wiry sort of limberness—like the quick men who work the log decks up in the Northwest. Theirs is the most dangerous job.

The ridges and intervening draws were like a tight series of triangles with the broader bases toward the rim of the canyon. They narrowed to sharp apexes as the slope gained momentum, then dead-ended in bigger draws. I made a point to hop from one to the next before I hit the end. I worked my way toward the avalanche chute that showed on the air photo so that I could have some vision of what I would be up against farther down.

A compass and a photo don't do you any good if you can't locate yourself on that photo. Shooting an azimuth in this kind of country can be the kiss of death if you have the kind of personality that prevents deviation from true lines of progress. It is like organized religion in that sense. The land must be read, courses altered, and a three dimensional kind of rhythm developed. It is a knowledge learned through a repetition of mistakes and good moves. I found myself grabbing the right sized saplings and swaying along from one to another, then boomeranging across a draw to a different little ridge and more steepness. My fantasy was that I was a primate swinging through the canopy of a Brazilian jungle.

The deer and elk had clawed trails through all of it that followed the contours impeccably. Several of them converged into a landing that watched over the avalanche chute, which turned out not to be a chute at all. It was a very deep, sharp meadowed canyon that pulled hundreds of smaller draws into it.

I don't know what the animals do at these places. No studies have been done on why the earth is pounded flat and worn at what is always the best views of wonderful places. Why is it that these airy spots seem to supply a meeting place of sorts that goes unexplored by science? The neatly piled hills of droppings tell you that a commerce occurs here. A deer must be standing still to fashion such a pile. It is fertile ground for some budding graduate student who is willing to hypothesize acceptable alternatives—the animals are scouting for danger, the animals are scouting for forage, the animals are scenting a prominent landmark to communicate with other herds, the animals are reincarnated Sung Dynasty poets. . . .

When we were young John Gierach and I fancied we were the natural heirs to Hanshan and Shih-te, the T'ang Dynasty laughing hermits of Cold Mountain. The story goes that they lived off herbs and berries and the pure water of the mountain with an occasional foray to the garbage pails behind the Kuo-ch'ing Temple where Shih-te was a cook. When a guy from town tried to corner them to ask about the road to enlightenment, they took off into a cave, which closed behind them and turned to rock. Their tracks on the trail vanished, but they had left poems scratched into rocks and on pieces of bark. The jist of it was that you should try to make it to Cold Mountain or at least take a hike.

John and I hung out in Fern Canyon, which is off the Mesa Trail outside Boulder, Colorado. There were no trails in the canyon, but bushwhacking was easy. We picked days when the clouds were low slung into the granite and fine drops of water would condense on the ends of our hair. You could work it so that you were in the clouds and rock and that was all. The rock radiated a warmth gathered from another day's sunshine and the clouds scudded by, opening and closing. Now and then we would catch a glimpse of a bonsai-like pine somehow managing to grow out from a crack in the rocks and we'd scream and giggle at it at the top of our lungs. I think we meant it, too. When we'd had enough we hiked back into town and ate hamburgers and got drunk. Sometimes we showed up later at Jack Collom's. He was a real, working poet who lived in Boulder.

I made my way down the crumbly edge between what was now a scrubby stand of aspen and the rim of the vertical canyon. The exposure was real but not in the sense of being on a sheer rock cliff. It was a feeling of unsure footing, the possibility that the earth itself might fold under me and send me cartwheeling in an uncontrollable tailspin down to the willows and stream below. It was an insidious, unarrestable kind of exposure that can occur when you are alone and off where no one would think to look for you. I edged closer to the trees.

You must keep to the high ground in these places to maintain options. I charted my route where I found it, but more often than not it was an intuitive process. With the country laid out and open before me I could go with the feel of it, letting my feet find the right contours. It has to come of experience and always some luck. By letting the information come in straight through my senses and out my feet I have often found the interdiction of logic to be uncalled for and unwanted. I save reason for the tight spots.

The rim eventually narrowed to a ridge and finally a point that called for a decision. Another side canyon had come in from the north and the two converged below me. I could see the floor of Rough Canyon about a quarter mile ahead through a jumble of brush and red earth. The point dropped off steeply through several hundred feet of scree and stunted pine.

I could backtrack up the ridge and cut across the intervening canyon up higher and in more safety, then come onto a series of rolling benches and flats that appeared to drop neatly to the bottom of the main canyon, *or* I could take off down the scree, fight my way through the willows in the bottom and hop onto game trail that led to the rim on the other side. Hanshan would have hunkered down on his butt, settled into the scree, and felt the wave of earth gather around him as he glided into the willows, cackling and giggling the whole way. . . .

I brushed the scree off my pants when I stood up at the willows. There was a rock outcropping that I skirted, then dropped into the brush. Working through heavy brush is like being caught in a political demonstration that has gotten unruly. There are a thousand hands and feet clawing at you trying to get away before they lob the tear gas into the crowd. It is far worse if panic occurs. You try not to go down underfoot, you try to stay calm. . .you aim for the light.

I made it across the stream and out of the willows. It was a short hop up to the game trail. I angled off toward the benches and floated down to the bottom of Rough Canyon. These flats were the place you would camp, the place you would come back to each morning after looking for the biggest bull elk of your life, the place where you would decide if you would kill him when you found him. I'd brew a cup of coffee for anybody that I ever met in the bottom of Rough Canyon.

The creek coming through it all was bigger than I thought and the sound filled the bottom. It was not full this late in the season, but I could see where it had been. Whole trees and parts of trees were stacked in the bends. Roots hung out of the banks where the soil had been carted away by high water. Farther up lighter debris hung from the lower branches of thick-trunked spruce. Some of them soared well over a hundred feet.

There was a stand of aspen on the other side of the creek and I hopped over the red cobbles sticking out of the water to get to it. The light was colored from reflection off the towering cliffs of red sandstone on the northeast side of the canyon. The cliffs were huge and spectacular and somehow complemented by the aspen that grew to the bottom of them. The broadleaf openness of the aspen draws you in as a place that could be habitable over the longer run of things. If you have ever been among them during a windstorm you know it isn't true, because they are shallow rooted and blow over easily and fatally, but the attraction still holds like so many other things that are beautiful and sometimes deadly.

I found a spot where a feeder stream washed over the sandstone. There were a few trees that had turned and the light was on them. I laid down in the warmth of it and took a nap because I knew I would have to come out of the canyon soon.

I am not sure how I end up alone in places like Rough Canyon, but I think it is because I don't know enough about women. I have noticed that my friends who know more stay closer to town. It isn't a gender thing because I know some women who don't know enough about men. There is just a group of us who are out there wandering around, poking ourselves into wild places, trying to find something that is gorgeous. We are dependent on the seasons in the broadest terms. There is more to it than a change in the weather.

Coming out of Rough Canyon was no more than a grunt. I power walked using my hands to pump my knees and met the contour head on as much as possible. When I tired, I side-hilled on a game trail until I was ready to power walk the slope again. I tacked between game trail and sweat. It did not seem to take as long as it did, but I still made it out of Rough Canyon on company time.

In the clearcut the Oshá was prime for harvest. It had gone to seed and the leaves were just yellowing from the nighttime cold. It is the herb that cures anything. The Spanish sheepherders say a bit of the root stuffed in a boot or tucked in your bedroll will scare off rattlers. You can use it for coughs, the flu, high blood pressure, indigestion. It can come in handy after births. It is a plant for all seasons.

I went to the truck and unsheathed the fire shovel. These tools are to be kept sharp and used only in the event of fire. The easiest Oshá to dig is the clumps that grow in the gravelly soil near the old skid roads. I worked them by driving the shovel deep and then levering the rootstocks up to the surface. They are big and gnarly and have an astringent stink to them. I finished my day with two garbage bags full of the roots and a not so sharp fire shovel. They would come in handy when dried and I might be able to sell or trade some down in New Mexico when I got laid off.

There is one catch to this Oshá business. The plant looks a little like poison hemlock. You shouldn't find hemlock at 11,000 feet, but the thought hovers in your mind like Hanshan's disappearance into the mountain or aspen in the wind or the biggest bull elk of your life. It is very much like a job that would pull you into the seasons of a place like Rough Canyon.

GLADE STATION

The structure is common where the snows come heavy. The foundation is a simple square, but it is the roof that distinguishes the building. It is a perfect pyramid sloped to slough off the winters. The chimney comes up through it and is just slightly higher than the apex of the roof. There is a covered porch facing west that is a good place to leave your boots when the weather is wet and they are stacked three inches in gumbo. It is also a good place to watch the long sunsets that come up out of the desert and are sometimes colored with the light from Monument Valley, which is a long distance to the south.

The Glade Guard Station was built in that great flurry of Depression-era activity that saw the Civilian Conservation Corps nailing studs and boards together throughout the isolated reaches of the forests. They had the time, and the buildings are sturdy. In those days forest districts were small and the rangers often came up to the stations with their families and spent the summers riding the range and timber, doing what rangers did in those days. There might have been a technician along, but that was it. When the snows came the ranger who occupied the Glade Station packed the kids and wife up and came into winter quarters at Cortez before the drifts piled up to the windowsills.

There is a long, flat meadow called the Glade that appears on the topographic maps. It begins near the base of Glade Mountain and runs west until it drops off steeply. There it becomes Glade Canyon and plunges 2000 feet down into the Dolores River canyon where the river begins to bend around Mountain Sheep Point. If you listen around you'll find that when someone is talking about the Glade they won't just be talking about what's on the map.

"The Glade" means the whole of that flattopped pine country that is as far west as the San Juan National Forest goes. I think of it as the country northwest of Salter Y, but others might say it even includes the Beaver Creek country west of McPhee Park. It certainly ends at Disappointment Valley up north and the Dolores River to the west. That is where it falls off into the plateau country of western Colorado and Utah and you can see the Abajos and La Sals floating like mountainous islands in the canyon country beyond the beanfields.

I came to the Dolores district off a season in the big timber country up in northern Idaho. It was a homecoming of sorts because I'd spent most of my time in Colorado. When you've been at it long enough you begin to meet up with people you've known from where you've been or people who know about where you are heading. I'd met a woman up in Idaho who'd just come up from the Dolores. When I came into the country at least I'd heard a little about it. We're all of us migratory.

I'd come onto the district late and the crew was already spiked out at the Glade Station. "Spiking out" was a term I'd learned from the Forest Service. The elk hunters used it, too. As far as I could tell it meant camping out or staying away in the woods overnight. I'd asked a few people how the word came about, but nobody knew for sure. Some thought it referred to tent spikes, others thought maybe the term had originated with railroad crews. The hunters thought maybe it had to do with a yearling or "spike" bull elk. Wherever the term came from, you will hear it now and then when in the offbeat places in the West.

I was looking forward to spiking at the Glade Station because I'd heard my old sidekick Kim was on the district. We'd met on the Pikes Peak ranger district and it was all right meeting up in a new place. Kim and I went back a long way. We'd worked district fires together and the big quick-running project blazes in California. Both of us knew the ins and outs of the timber game. I didn't know anybody who could outwork or outtalk Kim and she could go head to head with them all and drink them under the table. She'd thrown a party when I left for Idaho that became legendary on the Pike. The following day the entire crew drove quietly to an isolated meadow up Phantom Creek and we slept until two in the afternoon. Some people had taken to calling her Wild Child that year.

The Glade Station wasn't more than an hour and a half from the district office, but there was a sense of going far away when you took off for it. There is a remoteness woven into the country there that is a variation on wildness. Most of the huge yellowbark pines were logged off years ago. This isn't to say that there is no thickness to it, because there are still stands of big blackjack pine that grow faster than any I've seen in the state. A lot of it was logged selectively, at least initially, and if you didn't know what you were looking for you would be hard pressed to find the stumps. The clearcutting came later and you would have no trouble picking up on that.

The Glade has more to do with an intensity of plainness and regularity—tree after tree goes out to the edge and down into the canyons. The variations are subtle and need to be learned. The heat in the summer can be oppressive and you wonder how it is that a forest is here. Most of all it is not a spectacular place. You won't find anyone tooling over the gravel roads on a mountain bike or erecting a day-glo tent next to their internal framed backpack. The canyons that crash off of the flat down into the Dolores River stay empty except for maybe a cowboy looking for strays.

The environmentalists have written it off as permanently trashed. There is a fidgety lack of purity that borders on contamination at the Glade. You have to come to it balanced rather than seeking balance from it. This is not textbook wildness and it is frightening country for idealists. I am a fan of the places that lie in the shadows of the spectacular. They are good spots to improve your vision.

I talked with Kim late into the first night owing in no small measure to the beer she had brought along to camp. I'd managed to land in a gathering of sorts. Other members of the crew had invited their families up to spend the night and they'd all chipped in on the feast that spread across the table. Larry's wife was there and Dan Rael's wife had brought along their two little boys. Dan is quiet but rock solid. Gather his family around him and there is a contagious sense of wholeness.

I'm not sure if Cowboy Bob was there or not, but he lived in a camper close by. He was out riding for the Bradfield Ranch most of the time, but now and then somebody coaxed him over to the guard station. He was a big man who, in the early morning, you would occasionally hear arguing with his horses that were corralled at a reservoir below the house. He was up at dawn, all spurs and chaps, and back at dark every day with a tip of the hat and a wave. We never saw much of him. He said he wanted to have college so that he'd have the smarts to run a herd of his own someday. Cowboy Bob figured that cowboying somebody else's cows was no way to live.

Kim filled me in on the job. The crew was marking timber over a huge area. It was a very light mark — mostly a little thinning and taking out the high-risk trees. More often than not it was a matter of keeping from getting lost in the oak brush and picking ticks off yourself. It was flat and easygoing with plenty of walking and a break every now and then when Dan cruised a tree. We had to keep our eyes peeled for taller trees with interlocking crowns—these were going to be saved for the Abert's squirrels. All in all a pretty easy prescription compared to some of the marking guides I'd seen. The crew was on ten-hour days, which meant we had a four-day week and needed to spike for three nights.

A few beers later I got the story on the Pikes Peak district. Things had been getting seriously crazy when I'd left for Idaho and Kim said it all got out of control. Seasonal crews are fragile families and any number of things can set them to bickering. Our crew had simply fallen prey to what we call the "four year" rule. Any crew of seasonals together for four years or more will begin to either kill each other or start breeding, both of which can get hairy. The key is to get a job where you work alone or jump to another forest when the scenery gets too familiar. Most of the seasonals who last are always looking over the far hill anyway and three or four years in one place is about all they can stand. It's a trait that runs deep and implies that you had better be willing to resole the boots a few times.

The routine did not vary. We were up and out by seven and making the short drive to whatever section it was that we were working on. We got out, lined up, and set off marking. The Glade country is thick in oak and it was easy to end up in a patch and lose the flow of things, ending up way off in somebody else's area. Some of the lines were over a half mile long. We'd lay around when a cruise tree came up, then take off again. The heat came up a little before noon. It was a combination ponderosa pine and desert sort of heat. You wanted water but it didn't solve anything. It was the kind of heat that takes the color out of everything. I found myself watching the ground for horned toads. Early in the season they were small, maybe the size of a fifty cent piece, but as the season progressed they grew as large as my hand. They had a reddish color about where their ears would be if they had ears. If I was slow about it I could stoop down and pet them.

There is another presence out on the Glade. I don't know what tribes hunted that ground, but there were flakes of flint and churt everywhere. Sometimes we picked up perfect arrowheads from a newly washed area or came across painted shards of pottery. There was one spot that had been a camp and there was a heap of flakes where the hunters must have come year after year and worked rock into points. It was inescapable wherever we were that there had been people before us who might have been smart enough to lay up under the brush and chip points during the heat of the day. People who might have worked by clocks that fit into things better than the electronic boxes we had strapped to our wrists and spent too much time watching. The heat and the ghosts worked at keeping our heads down and I am sure that we missed some trees that should have been looked at.

The best times on the Glade were in the early morning and the evening. The country seeps into you then, but you don't know about it until it has circulated completely. That's when you become aware of something on the inside that is not apparent on the outside. Simple things go unreckoned, like the slant of the light on the grass or the birds gathered around a watering hole. There is a richness that is elusive and uncentered. Once I saw a doe and two very new fawns. I was furiously snapping pictures when it dawned on me to watch the light change on them. Just watch it, I thought. There is no better light than this.

Most often we came in tired. Somebody would cook dinner and we'd lie around or go out on the porch and climb into the sunset. You could wrap it around yourself in such a way that you were glad everyone was sitting around you, but they seemed so far away as to be out of range. After dark we fired the mantles on the gas lamps in each of the rooms and sat down to books or sometimes a game of cribbage. You will find a cribbage board in most every spike camp you come across. It is a game that ameliorates time spent waiting for whatever reason.

Kim and I decided to drive up to the top of Glade Mountain one evening because we knew we would be able to see a long way from there. Once there'd been a fire tower out on the point, but it had been taken down and moved over to the top of the next ridge to the east that was 200 feet higher. The road to the old pad was rough but still okay. You must get there when the sun is setting and go all the way out to the end of the road. Out beyond Ryman Creek, Cash Gulch, and the Hogback lies Disappointment Valley. There are a hundred alkali hills and rimrocked mesas that catch the light and throw it around a photon at a time. Up north is Lone Cone on the border of the Uncompahgre. It is a place where you point, arm straight out, and attempt identifications.

"Is that Brumley Point? Where does Disappointment Creek run? Those must be the Wilsons, can you see El Diente?" Out to the west are the mountain islands, the Abajos and La Sals, floating higher than they should on the desert haze. After the sun sets it turns inky and purple down low in the valley, but is still blue up high. When things start to become indistinguishable the nighthawks come out after insects. They carve incredible dives into the tail end of the light and you hear the rush of air whirring over their wings. When it's finally dark all you see is the flash of the white bars across their wings and you know that if light has sound, this is it.

We watched the lights come on out in Utah and across the valleys. Mostly they were ranch houses. Off in the distance we thought we made out Monticello or maybe Blanding. You wonder what it is that goes on out there. Who is drunk that night or what is out on the dinner table or maybe who's off with somebody else's sweetheart? It could be that the canopy of light and electricity over Monticello is no more or no less "natural" than the canopy of leaves over a grove of aspen out on the Glade. Distance does that.

We worked through the summer and into the autumn. Late in September the rain came every afternoon and we pulled our slickers on and kept moving. We stayed off the hills and laid low in the oak when the lightning was close. At night we dried our clothes. The rain got colder every day. One day I was off alone checking the timber up ahead. The rain was heavy and in sheets. I followed an old skid road back when I was done. Around a bend there was a bear sitting in the middle of the road, covered in mud, relishing a branch of raspberries. We looked at each other. I circled wide. . . .

The next day they called us back into town. We'd finish the sale the next year, they said. We cussed the weather and the Glade all the way into town. I fell off to sleep in the back seat and dreamed about the benefits of staying wide and being a bear.

THE TREE PLANTERS

We saw them from a very long distance off. I picked their car off as a model I remembered from my teenage years. A kind of phantom from 1965 when it was important for a kid from suburbia to know all the models and makes that were to clutter his future. This particular model I knew better than most because it was the same as the first car I had ever owned. The 1965 Plymouth Valiant came at the very end of that time when it was still possible to buy a new car at the birth of your child and maybe, just possibly, give it to him to junk around in his junior year of high school.

I'd bought mine from my dad in 1970 and banged it through the smoky haze of my college years and upon graduation into the busy workaday world of a minimum wage, grunt labor landscaper. The year before I bought it I'd borrowed it from the old man and driven to the Woodstock Music Festival in upstate New York. It had proven its value as a beast of burden when the festival ended and we piled twenty people in and on it and rumbled them out to their own cars hobbled in the congestion some ten miles out. The springs were arched backwards and we rubbed some rubber off the tires, but the Valiant made it and Dad never knew the difference. I sold it for fifty bucks when I left Boulder for Eagle's Nest, New Mexico, in 1975.

I watched the car wrap around the first switchback at the bottom of the clearcut, hoping that it had the optional 225-cubic-inch engine as opposed to the standard 190-inch model. This was an about average northern Idaho clearcut and it would be no easy chore for a standard engine to grind up through the remaining six or seven switchbacks to where we stood near the top. Depending on how a clearcut was slicked off and the desperation of the forester who had laid it out, a northern Idaho clearcut could go anywhere from almost level to a seventy percent slope.

The clearcut we were standing in had been harvested before the new wave thought that inspired the national Forest Management Act of 1976 and showed all the characteristics of the good ole days approach to logging. It was a big mother and had probably supplied logs in the tens of thousands and volume (the "Big V") in the millions of board feet.

This was my first day as a tree planting contract inspector. Tom, who was standing nearby, would be the main authority on the job. He had several seasons of the ins and outs of the planting business. I would serve as a helper and if things went okay maybe end up as numero uno inspector in some jobs coming up down the road. I'd been given a copy of the contract and the Reforestation Handbook.

My head was swimming with terms and conditions— plantable spots, J-roots, scalps, allowable distances, plot sizes, bareroot stock, containerized stock, air pockets, jelly rolling—the confusion aside, I was looking forward to the job because of overtime pay. Planting contracts always added up to big bucks for inspectors. Next to fire time it was about the only way to put something back against winter and unemployment.

The Valiant rounded the final switchback and pulled to a stop where Tom was standing. It was packed with gear and people. The longhair at the wheel rolled down the window and a cloud of marijuana smoke poured out.

"Say man, what's happening? Is that you, Tom?" He called through the smoke like he was off on a mountain in the distance.

Next to him sat the girl that I'd seen a thousand times around Boulder in the early 1970s. It wasn't the *exact* girl, of course, but it was the same girl—the long, straight hair, mostly blonde, blue jeans, and mountaineering boots. She had a flannel shirt on but I'd have bet she had a Mexican peasant blouse stashed somewhere for festive occasions. Odds were she didn't own a bra. Next to her sat a six- or seven-year-old boy with neatly trimmed hair that came down in bangs to his eyebrows. In the back seat there was a serious looking man. His moustache was meticulously trimmed and his hair cut short with none coming over the ears. He had round, thin, wire-rim glasses. There was also a stack of gear back there—tarps, wool jackets and shirts, raingear. The trunk of the Valiant was tied loosely over tents, food, and cooking paraphernalia. The springs of the car were flat out under the load.

The driver stepped out of the car and offered the joint to Tom. He shook his head no. The first of the tree planters had arrived and this was the meeting of the tribes. It was a meeting that made sense, unlike the one a few days earlier down in the district office—the prework conference. There we had all sat in a bored circle around the Contracting Officers Representative (COR) and gone over the nuts and bolts of the job. Acres, location of tree caches, who was responsible for what. The contractor got his licks in and we sat and nodded our heads.

Now the planters were coming. They had decided to camp on the timbered flat that was just outside of the clearcut. Tom knew the driver of the Valiant from other jobs and other years. They talked about how the planting had gone in Georgia over the winter, the weather, and who had gone where and the state of the gardens on a number of small acreages scattered throughout the Northwest.

These were the people who had bought up the cheap land and gone back to it the way we had all talked about ten years before. And they were still doing it. There would be about twenty more planters coming, they said. Some would be other New Age itinerants working for a grubstake, there might be an illegal alien or two, and always some folks just plainly and simply down on their luck.

There would be a camp just the other side of the seed-wall where the fires would be kept alive by the few people who ached too much to work or those who were taking care of the children. There is no harder work than planting trees, but there is enough money to be earned in the season to run a garden on or buy a drink. As inspectors we often wonder why we are on our side of the fence when we have so much more in common with the planters. The planters know this, too.

They started planting the next day. All of the crew hadn't arrived, but they would filter in. The contractor was there and anxious to get rolling. In most cases the contractor starts as a planter but sooner or later realizes that any real money to be made is by bidding the contracts. It is a high-risk endeavor. Planters may not show up or their work may not pass the inspection. The contracts are let in the spring or fall when the soil moisture is up from the rains or snow. Bad weather is the best for tree planting, but a spring snow can close things down and delay jobs that the contractor has lined up on another forest. A spring drought that reduces the soil moisture to unacceptable levels will cause the job to be cancelled. A contractor usually hires a foreman who actually runs the planting crews while he covers logistics or goes ahead to check out upcoming jobs. By and large, contractors as a group are an edgy bunch—like most businessmen.

The tools of the tree planter's trade are not complicated.

Although power augers and a type of digging bar called a dribel are occasionally used, the digging tool of choice and often necessity in the Rockies is the hoedad. It consists of a long blade about eighteen inches by five inches that is rounded to a point. The blade is bolted to the end of a stout, ax-like handle.

When a planter comes to a spot that is suitable, he first scalps the area free of sod and organic debris with the long side of the hoedad blade. The contracts call for the scalps to be in a twelve- to eighteen-inch radius around the tree to be planted. When the scalp is completed the pointed end of the hoedad is used to dig a narrow eight- or nine-inch hole to place the tree in. The dirt is backfilled in around the seedling and then tamped. A skilled planter can usually scalp the site with two or so strokes and create the hole with one power stroke where he drives the hoedad into the soil, then rocks it back and forth. The work is backbreaking.

A finer point of the planter's trade has to do with shade. There are species of trees like spruce that initially grow in the shade of another species like fir or aspen. In time the spruce, or climax species, will grow to dominate the more sun tolerant species and often eliminate them to form a pure stand of spruce.

In the succession of a forest the sun tolerant species are often the faster growers that come in after a fire or some other natural disturbance, while the sun intolerant species will grow somewhat more slowly under the shade and protection of the seral species. In a clearcut there is no overstory of the sun tolerant species and shade must be placed to protect the fragile seedlings and create a microclimate of suitable moisture and temperature.

This is accomplished when a planter locates the seedling he is planting so that it is shaded from the south and the west where the sun is most intense. To do this he finds a stump or the scattered debris from the logging of the clearcut and plants his tree to the north through east of it. In this way the seedling is shaded during the hottest parts of the day. Locating shade properly can be confusing when slopes and aspects are changing constantly.

Seedlings come from the nursery in two ways: bareroot or containerized. Bareroot stock comes with the roots exposed. They are usually "jelly-rolled" on or near the planting site in a slurry of root stimulator and vermiculite and rolled in burlap. The roots must be kept moist at all times. Containerized seedlings are the most commonly used. These are grown in a soil mixture in a small six- or eight-inch tube. Sometimes they come in "book packs" of six preformed square tubes of plastic. When the tree is planted it is removed from the bookpack or tube and the tree plus the soil around the roots is placed into the hole dug with the hoedad. Containerized trees seem more resistant to drying out than the bareroot type stock—at least under the actual conditions encountered during planting.

The science of it all was lost to the planters working their way through the clearcut that morning unless they were among the army of unemployed, trained foresters wandering the backcountry. The foreman had lined them out on the boundary of the clearcut and they worked their way across the top. It is a matter of how many trees he plants that determines a tree planter's day. They are usually paid by the tree. In the mountains of Idaho a really hot planter may get close to 1000 trees by working dawn to dusk, but a more likely average is 500. In the sandy, easy soil of the pine plantations of Georgia a winter planter may punch 3000 trees in where he doesn't have to worry about shade.

Tom and I took up our inspecting once the ten or so planters had some acres behind them. With only ten planters it is a piece of cake. We ran a few 1/100 acre plots and looked for scalps, trees planted too high, and dug some trees up to be sure that the holes hadn't been made too shallow and the seedling's roots folded back in the unforgivable shape of the J-root, which always kills the seedling.

The first few inspection plots on any contract are the most important. It is like the beginning of a big high school football game and you are on defense. It pays to give those guys on the other side a good shot or two just so they know that you are there. We tried to be fairly critical so that the contractor would know where we stood. This is a time when the players on both sides feel each other out. Before we were done with our second plot the contractor and the foreman came over to watch. These guys know that every inspector has his priorities when it comes to planting. Some inspectors watch for high trees; others are tough on scalps. If a foreman can figure what it is that the inspector looks for, he can keep that aspect of the job in line and maybe let other phases slip.

It always begins in a good-natured way with lots of "you bet we wills" and "Oh, yes, we're always careful about that," but the bottom line is that it is an adversary relationship and many contractors will do as little as possible to get a full payment on the job. The center of the entire deal is a type of scorecard the inspector keeps. This is a record of each plot, which usually includes the number of trees planted and if there were any planting errors. Through a series of calculations the inspector can come up with a percentage of trees planted correctly over the entire acreage. In most regions the planters need a ninety percent or better to receive a full payment for the job.

These are the cut and dry aspects of it, but the dynamics between the inspector and the planters are what really drive the work. After five or six twelve- or fourteen-hour days the real meat of the relationship starts to surface. The foreman usually begins the job by trying to put in as many "good" trees as possible early on so that he can build up a padding of high percentages. In this way if things go bad toward the end of the job, like planters getting pissed off and walking off the job, he will still have a high enough average to squeak through with the ninety percent.

We had trouble from the beginning with the scalps. The clearcut had a pretty good layer of sod, which made scalping tough on the planters. We gave the foreman a couple of warnings—inspectors are not supposed to talk to the crew members and are forbidden from giving direction to the planters—and he assured us things would change. When they didn't we started marking the crew off for scalps. This tends to shake things up in a hurry and when the foreman saw our sheet, which he checks periodically, he hit the ceiling and started screaming obscenities and jumping up and down. This is phase two of the relationship and usually the time when you hear about Forest Service trucks that had mysteriously blown up in the middle of the night and inspectors who don't have front teeth anymore. It is usually the phase where things straighten out once and for all or pandemonium breaks out. This is when talk about "claims" first comes up.

A claim is when the contractor takes the Forest Service to court and claims that there was some breach of contract. In many cases claims are settled out of court by the contracting officer in the supervisor's office. There are a few planting contractors who actually make the better part of their living off claims rather than planting, but they are in a minority.

Since contract inspectors, on the ground, don't really have any authority, most threats of claims are referred back to the district office. No one worries too much because the Forest Service writes the contracts to favor its own cause and the last thing most contractors want to do is get hung up in court with federal attorneys dragging their feet until the claimants run out of money. It is damn hard to beat Uncle Sam in court. So it goes on and the inspectors do their jobs and hope to get paid without having to get dental work or explain why the government rig blew up one evening.

The foreman on our job got the crew back on track and things settled down. When the percentages are high, the living is easy, as the planters on the hill say. We cranked our plots in methodically and the planters moved in that strange rhythm a crew has. It is like watching a flock of cranes work a field—one or two steps, then the hoedad swings into the scalp, then up again for the hole and place a tree in and tamp, then one or two more steps and repeat and repeat and repeat—dawn to dusk.

No matter what the rules, you get to know the planters. We talked with almost all of them here and there. It's usually pleasant and casual. We all knew what was at stake and the rules of the game, but there is more than work— we all work the winds of the seasons. Besides, some planting crews are known throughout entire regions for the parties they throw. Of course, the Forest Service is not supposed to fraternize with the coolies during a contract, but it is almost unavoidable, especially when you want to do it.

Tom and I ended up at the camp one night. He is a burly, black-bearded, big man with an earring. He plays an acoustic Martin guitar that can sound like wind in the timber. He sings and he likes women and parties. There was one that night. Somebody had just happened to bring up five or ten cases of beer and a couple gallons of wine. What could we say?

Some of the planters' parties I've made end up atrocities of drunkenness and loneliness. Women usually form a minor fraction of a crew and they are almost always with someone. We went just to look at them. Seeing a woman dancing off somewhere behind a truck with her old man is worth something when you are ten days into the woods.

Tom ended up with a circle of people around him singing and dancing. For a while I was off out of the way talking to a planter about places most people never see— canyons with steep walls, deserts far from water, and wild country. The kind of spots that trap you when you are there and when you come out stick in the mind. We end up with those places always hovering over our shoulders and waiting. The last thing I remember was dancing crazily around a fire. I woke up back in the Forest Service trailer and made work the next morning. The planters came on late.

The most legendary group of planters came out of Oregon and were called the Hoedads. I don't know if they are around anymore, but they were a fusion of all that

happened in the seventies. They were one of the first of
many co-op planting groups that applied the New Age
principles to the maximum. All it took was the rumor that
they were coming to plant on the district and the party
started.

These were the planters that *our* supervisors came out
to inspect. The men were strong, longhaired, and knew
about Buddha. Some of the women worked warm days
stripped to the waist, wearing only cut-off shorts and
mountain boots. This is what brought our bosses out—
sometimes even the district ranger—with their binoculars
and cameras. They watched from underneath their hard
hats with a squint in their eyes that furrowed the skin
clear back to the shaved hair around their ears. They always
left for the office to make quitting time and I think the
younger ones might have wondered on the long drive back
what it might be like to be a Hoedad. It was a thought
that was swept away when they sat back down at their
ordered office desks to work out next year's targets. I only
ever saw one of them at a Hoedad party and he didn't
stay a professional long—I heard he left his family and
bought a van. It is all what legends are made of, but the
parties were great—probably a lot like the mountain men's
rendezvous in the 1830s. And there were no police.

Our planters finished their job on time and passed. It
was a standard contract with little more than the standard
amount of bickering. The trees were in the ground okay.
I was a full-fledged inspector. Tom and I watched the last
of the planters' cars rattle off through the switchbacks and
burned some trash for warmth. They were off to a new
job like a band of gypsies.

I was given another contract to inspect that spring. The
planters were a group of Indians from a reservation down
south. About the sixth day into the job the men took off
after a moose that had wandered into the clearcut. After a
couple of days the women and children and a few men in
the camp packed up and headed off to see how the camas
were coming along in the Musselshell Meadow. We defaulted

the contract on them. Later that fall I found a fine crop of marijuana that obscured the stones that had made the fire ring in the center of their camp.

The years have gone by and I don't know how many planting jobs I've inspected. Most of them have been down in southwest Colorado where they found out too late that they couldn't get trees to grow on all the acres that were butched during the great clearcutting era. The clearcuts have been planted season after season only to fail. Now they are thinking mother nature might better take over and are letting some of the acres go to natural regeneration and the succession of species. It could be 400 years. . . .

I met Mary on a vast flat of ponderosa pine on the very edge of the forest in southwest Colorado. This is where the timber goes into the canyons and turns to pinion pine. It was cut heavily years ago and the subsequent plantations have by and large failed. She was the foreman of a crew and won me over when I got on her case about one of her planters who was putting two trees into a hole—not only was it against the rules, but it also meant the planter could double his count when payment time came. I didn't know who it was, but Mary went up to the trees in question and studied the boot marks in the moist soil. Next thing I knew she had the whole crew lined up and was looking at the bottoms of their boots. She found her man and I left when the shit hit the fan. I heard her screaming from a distance.

Later on in the job we got to talking. She was a stocky, dark woman with large breasts that rose perfectly from her chest without support. It was perfection to the point that I wondered what it was that held them out there in defiance of gravity. It could only have been genetics and years of swinging the hoedad. Planters develop tremendous upper body strength, especially in the pectorals.

It turns out that she had been to the Woodstock Music Festival in 1969. I ended up stopping by her tent when I saw her out there in the early mornings. We usually had a couple of beers before I had to hit the clearcut. There were 400,000 or so people at Woodstock and that is a lot, but they are scattered about now and the festival doesn't often come up unless you are in a planting camp. All of us are veterans in a sense. . . .

Mary is a contractor now. Being a woman placed her in a minority group that the government favors when contracts are let. She joined up with some other "investors" and they pretty much call the shots in their neck of the woods. She also knows the business, having come up from planter to where she is now. I saw her a year or so ago and she said she couldn't believe how much money she can make just be being what she is—$40,000 the first year. She said they partied for a month in the winter at high-class hotels where it was warm. No more tents.

I haven't seen her since then, but I hear about her from planters on some of the jobs. She is doing well, they say— bids most of the jobs because she knows the country. I could work for her anytime I want, but it is a hard life when you are used to government work. They say she has slimmed down and looks good, but I know that she would still sit down in the dirt and talk even though she has good clothes now. It's just the kind of person she is.

I would like to do that off in some clearcut with the planters ranging around us like a flock of cranes and the dogs running back and forth like sparrows in their wake. Maybe we could talk about what it was that happened at Woodstock one more time.

MOUNTAIN RUNNING

Chas Clifton and I had history on the Conejos River. We'd been meeting there almost every year for the past ten. We came with our wives, usually in July or September, and spent our time at the Menkhaven Lodge. Neither of the women cared much for fishing, so they usually wheeled down to Taos or took off hiking to places like Waterdog Lake or up along Elk Creek. At night, in the little fishing cabins, we'd all meet together and sometimes talk about the dreamy places that were farther back. The sound of them—Laguna Ruybal, Canon Escondido, El Rito Azul landed us on the border with New Mexico more easily than a road map.

Chas and I had never fished the Lake Fork, a smaller tributary of the Conejos, mainly because it required a bit of a hike and the big river itself had kept us occupied with its larger trout. However, there was a draw to the Lake Fork that was as simple as something that had always been there—Rio Grande cutthroat trout. These had once been the only trout in the whole of the Rio Grande drainage and they were the remembrance of a time before the introduction of brown, brook, and rainbow trout to Colorado. I hadn't come from Colorado either and I wanted to get my hands on something native or maybe more precisely something that had always been where it was supposed to be. So we went.

We managed to catch a few of the small, speckled natives and let them go, then headed back down the trail to the truck. When we got there I realized that I'd forgotten my wading boots. They were sitting under a bush where I'd ditched them when we found that the Lake Fork was such a thin thread we wouldn't need to wade. I decided to run back up and get them while Chas drove up the gravelled road to check out the old mining town of Platoro.

The point is that I decided to run, and I remember the look Chas gave me. He came from Forest Service stock and had actually been born up in Del Norte when his dad was a ranger there. Now, in the Forest Service and pretty much throughout the West, you did not run. If possible the idea was to drive in two-wheel, if not that, four-wheel, and then take to horseback if it was still too rough. All other options having failed, then it was time to throw a pack on and hike, but by God you hiked, because a man at least had to have his dignity. He only ran if there was wildfire licking up his ass and only then if there wasn't time to turn and shoot at it.

I took off at a jog. There had been some great exceptions to the no-run ethic in the West that I admired, most notably John Colter and the two hundred or so miles of running away from the Indians he accomplished in the Yellowstone country. It wasn't Colter that I admired, he was simply running for his life, but the native Americans that chased him. Almost to a tribe the native Americans have all been and still are great cross-country runners.

Running trails is not like pounding over the asphalt in town. There is life to the turns and bends and roots and rocks. It is no place for absolute speed. I tried to go fast when the going was fast and do the best I could when it wasn't. There's a pace to this kind of running that reminds me of the way that certain birds fly. They take the air on in a series of crests and troughs, almost like the swells on the ocean. I have watched them glide quickly down into the trough, then turn up and rise almost to the point of stalling, then turn down and glide again into the next trough. I felt that way running across the contours along the Lake Fork—almost resting coming down the hills, then coasting up to the top of the next rise, then drifting slowly over and gliding down the next hill.

I made it down the mountain before Chas got back to our meeting place and headed up the main road toward Platoro. I had in my mind that I'd run the whole way to Platoro, find him moseying around somewhere, and ask what the holdup was, but he came around the bend before long. I was sweating when I hopped into the truck and we drove back to the lodge without talking about running except a poke here and there about why a man would do something like that. Actually, I think Chas was kind of behind me, at least in spirit, because where he came from it was sort of revolutionary and I know Chas to have leanings toward things of that nature. We wouldn't get along if he didn't.

I can't ever remember not running nor can I remember ever trying to be *just* a runner when I grew up. It would be hard to see an affinity for it in my familiy with the possible exception of my mother. She didn't run but had the toughness of mind that I have come to appreciate in distance runners. Most of my running when I was younger was against the really tough competitors, like a night of too much drinking, or too many cigarettes. For me it was more a matter of getting something *out* of my system than getting into shape or even staying in shape. I kept the score even as best I could.

The idea of running played into the life of a seasonal with the Forest Service. Seasonals are the major source of grunts for firefighting duty and they were required to pass a physical fitness test if they wanted to get in on fireline duty. The test was called the step test and was as simple as stepping up and down on a bench for five minutes, then having your pulse taken. Depending on weight and age the score you got either qualified you for the line or not.

A number of seemingly fit people had trouble with the test, and the alternative was to run one and a half miles in a certain time that depended on the altitude of the place where the test was run. I could dash through the run with no trouble at all but sometimes had problems with the step test. If I trained up by running distance for a few weeks before the step test I could usually pass it, too.

The step test was important to us because the Forest Service, all through the ranks, prided itself on physical prowess. There is a tradition of very tough characters throughout its history, and with the exception of a new breed of lard-ass bureaucrats, that tradition still holds. For us seasonals there was even more to it—many of our job descriptions *required* that we pass the step test.

Aside from the step test I took to running naturally when I came on board with the Forest Service. When I started off doing timber exams I got into the habit of jogging between the stands of timber to save time. Most often it was as simple as finding a decent game trail and loping off in the general direction of where I was supposed to be with a periodic stop to check the aerial photographs and shoot an azimuth.

These were the runs where I began to learn how to make it through the forest with speed. There were tricks like a little stutter step coming downhill on rough ground that helped avoid catching a toe on a rock or root, the bobbing and weaving to avoid low hanging branches, and what I came to learn was called power walking. Power walking came into play when a hill was just too steep to jog up. In these situations I walked fast by pushing down on one knee, then the other to add spunk to my flight up the hill.

Distance running is actually pretty cerebral and I learned that after some distance I could exchange one mind for another. I would start off aware of all the things that are usual—a goodbye kiss from my wife, an instruction from the boss, or a bill I hadn't paid, but as I gained distance over the ground I seemed to gain distance from the usual, too. I managed to become an observer of myself—I saw my feet stuttering to avoid a rock or my head bobbing to miss a branch. The unpaid bill was still there, but my connection to it blurred.

All these things still rattled around in my mind, but I was able to connect with another more native part of it, a part that is always there, watching and thinking but not necessarily letting me know what's going on. It isn't something I would get too Zen about, just another way to go distance. Sometimes, if I went very long, I tried to imagine myself as the deer or the elk on that trail and felt the wind slide over me.

I had some years where I took the running seriously. It began when I quit chewing tobacco and ran to forget about it. I decided that I would run in some of the 10,000-meter foot races that were popular around the Four Corners. I bought a slinky pair of running shorts and a pair of New Balance 565 running shoes. The 565s were more than just any old pair of jogging shoes. They had rubber cleated bottoms that were made for off pavement travel and represented what I had in the back of my mind. I would be a mountain runner, which is the craziest of all runners.

It took a year just to get into the swing of it. In the summer I ran every night after work. We were usually in spike camp, most often at elevation in the spruce country, so I roamed the old logging roads. I started at a few miles and worked it up to six or seven a few nights a week.

The crew, again in that great western tradition, thought I was nuts. Each afternoon when we came in from a day in the field, when they were ready for an ice cold one, I climbed into my gear and took off down the road. I ran, cut most of the meat from my diet, and came back and went to bed early. By the winter my weight was coming down to a runner's weight.

I kept running over the winter. The act of just going out to jog in the snow and cold didn't do much in the way of conditioning, but I figured it would toughen up my thinking. By the next spring my weight was down from close to 190 to 175, which is light on a six-foot-four-inch frame. I began to look like a prisoner of war, which fit me in nicely with the other mountain runners I knew. There were some other funny side effects, like a forty-five beats per minute resting heart rate.

By the next spring I had a plan. I figured I'd run in a 10K race once a month while working myself up for the Kendall Mountain Run and the Imogene Pass Run. I ran my first race during the Bear Dance festivities at the Southern Ute Indian Reservation. I thought making a circuit of the various reservation foot races might be interesting and the fact was I still took the native Americans as the experts on the subject. When I told my reservation running plans to Don Hoffheins, a running friend at the Forest Service, he said good luck—they'll eat you alive. Don had grown up in Albuquerque and run cross country in high school. Many of their races were at the neighboring reservations and his memories were still vivid of the beatings they took.

I was a little nervous about the whole thing and decided to pick another runner my age to follow. I didn't think he looked any different than me, but after about two miles he was tearing me up. I dropped back, spent, and jogged in for the rest of the race. I had been totally wasted after the first few miles and learned for the first time that having other runners around can cause funny things to happen. It became apparent that I should *save* something for the end of the race.

When the final runners came across the line we were all gathered into a big circle and did some sort of victory dance to the sound of beating drums and a singing medicine man. The great Indian runner Al Waquie had been in the race as a kind of honored guest and paced the field by a couple of minutes—he hadn't even breathed hard. After the dancing he talked to us about his own running and the long tradition of distance running on the Jemez Pueblo. He'd come from a family of runners that went back generations.

I hadn't been much for speed, but I'd made my first race and somehow in that circle afterwards felt connected to the great native runners of the land. They gave us tee-shirts and to this day if I wear it on the reservation people's hands wave up in salutation. Now I know the Indian runners by their lack of noise. They are the quietest runners around, no shoes slapping the asphalt or trail, and once you've been around for a while you come to realize that this is the sound of the very best distance men.

I ran a race over in Dolores once during Escalante Days. It was a hilly course that started off hard. By that time I had come to understand that finishing in the middle of the pack was my station in the running world. I took the course as best I could and began to add a little speed with a mile or so left. My mind had opened from the exhaustion and somewhere in it I heard a quiet pitty-pat, pitty-pat, pitty-pat. I knew that sound and looked over my shoulder to see who was coming up on me but couldn't see anyone. It all happened in a matter of seconds, but the sound haunted me until a nine-year-old Navajo, who came up about to my waist, passed me fast on the inside—I'd looked right over his head when I glanced back!

The Kendall Mountain Run is a kind of home-brewed community event that starts in Silverton, Colorado. Although known as one of the tougher mountain foot races in Colorado, publicity is kept to a minimum and many of the runners are from the Four Corners area. This doesn't mean the word doesn't get out, it's just not at the sort of level to be found at the really high rolling races like the Pikes Peak Marathon. Everybody likes it that way.

The run usually takes place toward the end of July when the snow lets up its hold on the mountain country around Silverton. The runners start showing up around seven in the morning to register at Smedley's Ice Cream

Parlor. The top of Kendall Mountain is visible from town and I stole some glances at it after I registered. The run is thirteen miles, which is not a lot in the world of distance running, but the distinction at Kendall is the elevation and rate of climb. Silverton is at 9280 feet where the race starts. The summit of Kendall is at about 13,066 feet. This means that the runners cover about 3800 *vertical* feet in the six and a half miles to the top. That is intense.

Mountain runners are a breed apart from their everyday flatter running counterparts at lower elevations. I noticed ski bums, a few old longhairs, and just general duty mountain folk. Some of them were wearing cut-off jeans. Another contingent was the high tech folks who wore the latest polypropylene outfits and carried fanny packs with extra clothes. The grade coming *down* from Kendall is so severe and so strewn with rock and gravel that some runners wore knee pads in hopes of breaking what seemed to be the inevitable falls and crashes that can happen when legs are wasted from the run up.

Around eight an official gathered us at the starting line in front of the Grand Imperial Hotel. There was a short talk. They wanted us to know that there wouldn't be any judges on the summit if the lightning got ugly. There would be three aid stations going up and two coming down to supply water and candy bars. It was that simple. The race started.

The run up Kendall Mountain does not stay flat for long. We crossed the Animas River right out of Silverton and headed up an old jeep road that comes within 250 vertical feet of the summit—from there it's a scramble through the rocks, scree, and a little meadow to the summit. You must touch the pole up there to be official. I started at an easy jog on the first hill. I wanted to be able to ease along without going into a walk for as long as possible. I lasted maybe one-third or half a mile before I was crushed into a half walk.

I didn't stay in my mind for long. I thought about anything but the run, which was relentlessly up, and went between a fast walk, slower walk, and power walk with a little jogging where I could. Everybody else in my particular knot of runners was doing the same. The leaders were way out ahead. I remember few other parts of it. I saw the leaders on their way down when I still had forty minutes yet to the summit and we cheered them. What I do remember is when the jeep road ended. The route from there is straight up, hand over hand, through the rocks. Now and then someone up ahead kicked one loose and we dodged and yelled. At the top I touched the pole and drank a glass of kool-aid.

The trip down was brutal. With our legs shot from the trip up, putting the brakes on in the steeper portions was difficult. I saw runners career out of control and crash into heaps. One guy in front of me sprained his ankle, got up, ran fifty feet, fell again, then hobbled up and on. There is a beauty to the surge of endorphins that puts the pain at a distance.

Finally, I could see Silverton. I passed some kids who were lined out on the road and cheering. I remember looking at them like they were some sort of show on television because there was so much distance physiologically between us although there were only three feet separating us. I thought I could see every hair on their heads. I tried to put a spurt on for the finish but nothing happened. I was done.

They had kegs of beer at the finish. I had four or five glasses with absolutely no effect. I'd come in an hour after the winners—eighty-eighth place, two hours, thirty-five minutes, and thirty seconds. Maybe no big deal, but I have the memory of it in my body, which is different from any thinking I have ever done—the next time I ran Kendall it wasn't so tough.

After Kendall Mountain I started running more than I'd ever run in my life. We were working a timber sale off of Roaring Forks Creek near the Orphan Butte country. Our camp was up high on a knoll above the creek and the gravel road that went by it made for good running. I didn't bust a gut but I managed to get in five to ten miles a night, which included a short stop now and then. It was August and the shaggy manes were busting up along the edge of the road.

I carried a bag with me and stopped to pick them when I could. Back at camp I sauteed them up with my dinner. As mushrooms go it has been said that the shaggy manes don't rate that high, but I like them if only for their ephemerality. They must be eaten quickly after picking or they deteriorate into a slimy mush. All this means is that they'll never be found in the produce section of the local grocery store, and that in itself is a lesson about wildness.

On the weekends when I was home in Durango I made long runs up Horse Gulch and along the Florida River every other week. I cached water at the halfway point and tried for twenty-two or twenty-three miles. Horse Gulch was a wonderful place for this. It began just outside of town and climbed steeply through a canyon that was strewn with the junk people had dumped there. Old cars, refrigerators, washing machines, dryers, and stoves littered the arroyos. Each appliance was full of bullet holes. It seemed like it wasn't enough that whatever package of technology laid out there was dead, it had been necessary to put bullets through the heart of it. You don't want a wounded washing machine tumbling into town and committing some atrocity on your underwear.

Once I topped out of the canyon the country flattened into a broad valley of pinion pine and juniper. In the late summer the dirt road was thick with the small-headed, wild kind of sunflowers that come up about chest high. By staying to the beaten down tire tracks I avoided being slapped silly by the drooping yellow heads. More than any place I had run, Horse Gulch put me at ease. I could drift into a place where I wasn't thinking at all but just coasting through that dry world. I could understand that something was coming into me but it wasn't through the normal channels, and I sometimes got the feeling that I was getting the *whole* thing at once.

On the long runs I tried to leave right at sunrise to avoid the August heat. I was coming back toward town through Horse Gulch on a Saturday after about fifteen miles when I came across a small band of elk. It was made up of all cows and their spring-born calves. I would have never looked for them there in the pinion but rather expected to see them in the aspen or spruce of the higher country. They'd been bedded down, then heard me, and rose up like one and took flight. Those animals glided, fog-like in their smoothness over the grass that morning. They could have been illusions, phantoms of one too many miles, but I crossed their tracks on the dusty road. They had been there and I took it as a sign.

My wife, Monica, and I took off for Ouray, Colorado, the starting point of the Imogene Pass Run, the afternoon before the race. We chugged up past Silverton and Kendall Mountain and over Red Mountain Pass down into the town. I'd made reservations at St. Elmo's, a snazzy hotel/restaurant that was offering a big carbo-load spaghetti dinner the night before the race.

The Imogene Pass Run is usually held around the first weekend in September. It is a classic run over the pass from Ouray to Telluride, Colorado. One of the reasons Monica had come was to drive the car around to Telluride, the long way, and meet me there after the eighteen-mile run. The run had started informally when another great mountain runner, Rick Trujillo, a five-time Pikes Peak Marathon winner, had organized a "little" run between his friends. The word got around and it turned into a kind of regional event for hard-ass runners.

The application form told the story: "Imogene Pass is exceptionally beautiful, even by Colorado standards, but the possibility of injury/hypothermia must be emphasized. The enclosed map and course description speak to the difficulty of the endeavor. Race physician(s), EMT's/mountain rescue teams, and other assistance will be provided, but runners bear the ultimate responsibility for their own safety. TRAIN yourself to negotiate very steep, rocky, and

(possibly) icy, snow covered ascents and descents. PRE-
PARE yourself to contend with hostile weather conditions—
bring at least a windbreaker, gloves, stocking cap, and
emergency blanket. If you like mountain running and are
prepared, you will be ecstatic over Imogene."

The map showed a course starting in Ouray at 7811
feet and travelling steadily up for eleven miles to Imogene
Pass at 13,114 feet. From there it blasted down to Telluride at 8745 feet in seven miles. It sounded sporting, but I
didn't sleep well the night before.

It was drizzling the morning of the race. We tried to
calm each other by saying that it was clear up on the pass,
but the odds were it was snowing. In 1982 there had been
thirteen inches. The run that year had been legendary. I
decided at the last minute to carry a fanny pack with some
sweat pants, gloves, windbreaker, and hat and ran back to
the hotel to get it, figuring the extra weight might build
character. At the starting line I paced around with Monica
until we ran into Ernst Baer, a friend and superlative
runner from Durango. We talked awhile until he noticed
something strung through my shoelaces.

"What are the bones in your shoes?" he said.

"Elk teeth. Buglers," I replied. He looked at me
strangely and went off to talk with someone else. I had
come to trust things like elk teeth a long time ago.

The starter called us to the line and I said goodbye to
Monica. He ran down a list of do's and don'ts, talked a
little about the weather, and started the race by saying,
"This is the party you came to. . . ."

The beginning of a distance run is always the hardest.
I'm usually all jazzed up and spend the first mile thinking
how stupid it all is. When the edge goes off and I'm settled
and let things go I forget about it and just run. It took a
little longer for me to calm down on Imogene. We hadn't
run very far before the snow began to mix in with the
rain. At Camp Bird, a little mining town five miles up the
road, it had turned completely to snow, but it wasn't
sticking to the ground. Some runners began turning back

before we made it to Upper Camp Bird, which lies in a high basin eight miles from the start. The jeep road out of Upper Camp Bird climbs steeply to timber line and that is where it got ugly. Those three miles to the summit go through 1800 vertical feet, and the snow was coming in hard. Somewhere in there it flattens into a false summit and on that day a blizzard. The temperature was twenty degrees with a twenty-two mile per hour wind driving the snow into the runners. Some of them went down with hypothermia. One case was so severe that the runner was saved from death only when another runner, who was jogging along with his burro, picked the guy up, threw him on the burro, and took him to a first aid station.

Near the summit Rick Trujillo was yelling encouragement to the rest of us. EMTs asked how we were to see if anyone was delirious from hypothermia. I'd managed to get my gloves and hat on along with the windbreaker, but my hands were too cold to get the sweat pants on. A runner next to me screamed out, "I'm from L.A. and I hate this shit!"

At the summit I grabbed a few candy bars and got the hell off of it. My eyeglasses were fogging badly then and I had to take them off, which for me is a minor crisis. I ran it on instruments over the rugged mile or so down to Tomboy Townsite. The weather began to clear a little there and I took time to gobble down the candy. The sugar kicked me right back into a jog.

I can't say where the snow turned back into rain, but I remember running into a wall of warm air. It was like splashing into a hot spring and I knew I would make it. I could see Telluride down below and the condition of the road improved to a good running surface. Nearer to town people lined the road and cheered. They have some kind of deal there where they radio your number ahead to the finish line and they announce your name as you cross. I had saved something the entire race and put on some speed coming in. I saw Monica at the finish line and could hear her yelling above all the others. Her voice brought me in.

I came in 100th at three hours, twenty-six minutes, and three seconds. An hour later, to the minute, of the winner. But everyone who comes over Imogene Pass on foot, as fast as he can go, wins—it's that sort of run.

Monica looked at me a little later and said she had expected me to be among the first group. I laughed, figuring that she was fooling around. There had never been a chance. I knew that and had even told her. But she was really serious. And I take that as more than a compliment.

I don't run as much anymore. I might put in a couple of miles here and there—maybe fifteen or twenty a week. Sometimes I'll break into a trot on a good game trail, particularly in the aspen, but I'm not competing much now. I get up Horse Gulch now and then, too. I don't think I ever really did compete in the true sense, unless it was to look for something that had always been there or something native or maybe just something that was where it was always supposed to be. Running alone, on the trails—it's still the only real kick.

GLORIOUS MISERY

Nipple Mountain rises about 700 feet from a broad bench that falls off steeply for 1000 feet to the West Dolores River. The mountain is little more than an oak and aspen covered knoll; but from the right angle, in the right light, the view from the areola-like bench brings a twinkle to men's eyes. The bench itself has evolved through a number of names. The 7½ minute topographic map out of the U.S. Geological Survey calls it Hooray Flats. I've heard it called Ouray Flats, probably after the Ute chief whose name has found its way to a number of places on the anatomy of that country. The Forest Service map, in what appears to be a move toward topographical dignification, calls it Hugh Ray Flats.

The logical guess is that some scrappy prospector climbed up from the river one day and found something there that made him shout "Hooray!" It might also have been that he shouted "Hooray!" because it was finally flat.

I like to think that he came up onto the flats at that perfect angle and with good light and stared up at the Nipple and sang out "Hooray!" When visions like this begin rising from the landscape, it figures into a kind of calendar where you've been out a month too long.

We made a spike camp on Hooray Flats the year that we surveyed the cutting blocks for the Groundhog Aspen Timber Sale. The road was too rough to haul trailers on, which were what we normally used when we camped out, so we put up a classic camp of big white canvas sheep-herder tents with small woodburning stoves in them. We had a deep, open firepit at the center of the camp.

Each day we went out in twos or alone to either traverse or cruise blocks of timber, and that country began to come into us. The big flat and smaller benches of aspen that stair-stepped down to the river were the calving grounds for a large herd of elk that migrated up from the McPhee country, then along House Creek, and finally to the Nipple. We followed their trails through the maze of aspen and underbrush because they were the easiest way around to the work. Now and then we spooked up a tribe of resting cows and calves that took off in a storm of hooves, barks, grunts, and shrieks.

The weather around the Nipple that summer was mostly fair, but when it turned, it turned with a vengeance. We watched it most closely toward the weekends because the only way to drive out of Hooray Flats was to skirt the Nipple on the north, then head past Dutchman Lake toward Groundhog Reservoir. The road was little more than a trace over grass and dirt. If a big thunderboomer with a lot of water in it came up over us and slicked up the hill there was no chance of driving out even with chains on all four tires, four-wheel drive, and winches. Stormy weather on Friday meant a walk down the Goble trail to the West Dolores and a phone call to town for a ride.

Toward the end of the job Dean Dallman and I ran a survey on a block of aspen that was flagged out on a sharp ridge that ran south of the Nipple. It was such steep going to get to it that the logic was we could leave it until the end in the hopes that we'd get pulled off the sale to go to a fire, but our time had run out and we trudged up its flank. The view from the ridge was a good one that took in the flats around Green Reservoir, the King Ranch up north, and the big slide of country down into Cottonwood Creek.

We were about halfway into the endless chain of azimuths, distances, and slope corrections that serve to corral a wild piece of land into sellable acres of timber when we caught sight of a tremendous storm boiling in from Beaver Mountain to the west. The thunderheads were typical in the most severe way, which means that the undersides were light, almost snowy in color—a sure sign that a curtain of hail would sweep in with the lightning. The thing was moving quickly behind a veil of virga and the clatter of the thunder echoed through the canyons.

The reasonable thing to do would have been to come off the ridge. In fact, these were the kind of things we talked about in our weekly safety meetings. Number 13, in italics, under the lightning section of the Health and Safety Code clearly stated, "Avoid tops of ridges, hilltops, wide open spaces, ledges. . . ." We decided that we could beat the storm and finish the traverse and that would save a walk back up the ridge. This, of course, would have been frowned upon by the always safety conscious boys back at the office who never climb steep ridges.

I may have swayed the decision. In his younger years my dad had been a weatherman and he instilled in me an affinity for ugly weather. We would wait together, faces pressed against the windowpanes or heads tilted up toward the sky, for any signs that the atmosphere might rip open. We scrutinized each front, low pressure trough, and ridge of air for any violence that might be hidden within. When it did come we savored it like fine wine, commenting quietly on the finer points of a clap of thunder, fierceness of a blizzard, or utter hugeness of the occasional edge of a hurricane that touched us as it ran up the East Coast.

It wasn't a reckless fascination, we always took cover, but more a wonderment at how big things can get in the world. Now and then some old friend of Dad's would blow in and they'd include me in on their stories about flying into the eye of a hurricane or watching tornados rip through Kansas. While some households held a cherished auto-graphed copy of a current novel, our bookcase had a special spot for a signed edition of *The Nature of Violent Storms* by Louis Battan.

The intensity of the storm that crashed into the Nipple country drove us from the ridge in a hurry. It came on a wind that ripped the shallow-rooted aspen from the ground. The hail that followed beat the leaves from the trees that were left standing. Lightning crashed so closely that the charge crackled the air. We raced off the high ground, slid-ing and falling on the hail-slicked leaves, toward the safety of the saddle between the ridge and the Nipple. When we finally made it, the rain was coming in sheets between the bolts of lightning that were landing all around us.

The storm was so loud that we yelled to each other, our faces a foot apart, to be heard. There, hunkered down in the saddle, soaking wet and cold, would have been the place to smoke a cigarette the way Dad used to when things got a little beyond good.

The other great storm of that season of camps on Hooray Flats came when big Jim fell hard in love with Suzanne. He was on the crew for the summer out from forestry school in Wisconsin. She was up from Flagstaff and the forestry school there. Along with Dean, who was also from Wisconsin, the three of them made up the younger contingent of students that drift onto the crews during the summer for experience. The rest of us, Larry, Randy, and I, were the old married men who came back year after year. We watched that romance the same way that we watched for the grass to green up in the spring or stuck our heads out of the tents in the morning to check the weather. Each of us alone, preparing for the coming season or day in the way that we saw things shaping up.

The beginning was the same as most beginnings of that kind and it charged us with the excitement that was left over when the two of them couldn't hold it in anymore. Theirs was a very civilized romance; they reserved any passion for quiet moments alone and after work or the weekends. On the job they were just part of the crew, figuring we knew nothing about them, which tickled us even more. These things rate right up there with the strongest weather and you can find yourself reaching for your poncho even if the storm misses you.

It went on with a wonderful delicacy for the better part of the summer until Suzanne showed up with her boyfriend from Arizona at the Escalante Days celebration in Dolores. Big Jim showed up, too, and got his heart broken to pieces. He didn't flare into any violence, which would have been awesome considering his size and stamina, but went the other way, inside of himself. He came to work the next Monday dragging and the great excitement in our camp reversed polarity and we turned our thinking to autumn and the hunting seasons. It wasn't but a few days before frost sealed off the ground on the flats and up on the Nipple.

Black Mesa lies about eight miles to the northeast of the Nipple as the crow flies. It rises steeply from Fish Creek and Little Fish Creek to a three-mile-square and somewhat flattened summit at a little over 11,000 feet. The name probably came from the dark, thick stands of Engelmann spruce and subalpine fir that once occupied the mesa top before they were clearcut out during the frenzied cutting years of the 1970s. Now that high country is a jumble of old logging roads and plantations of spruce that don't seem to have grown very well.

The elk like Black Mesa and there are people who believe that affinity is due to the clearcuts that opened it up. My hunch is that the elk were probably spending time on the mesa long before the clearcuts. It is an airy place with fine views from the edges and for connoisseurs of the weather, which elk by design or purpose seem to be, it is a magnificent theater of all the atmosphere has to offer. The Black Mesa is uniquely situated. It is southwest of a chain of rugged peaks that make up the Lizard Head Wilderness. In a weather sense it stands as a sentinel to the Rocky Mountains in its part of the world. The key to its vulnerability is Disappointment Valley, which lies to the west and just a hair north of the mesa. The valley provides a straight shot for any storm that has picked up momentum over the deserts and canyons to the west. The first obstacle that they hammer into is the Black Mesa.

I occasionally had business on Black Mesa. In the springtime there was the ritual of tree planting. Later in the season we sometimes offered up a few trees here and there for salvage. Almost every year we conducted some sort of survey on past tree plantings and how they fared against a population explosion of pocket gophers. The results tended to not be promising.

In the process of the work I came to know that spring always came a couple of months late to the mesa because of the huge amount of snow that dumped on it. In the summer that place was a lightning rod where we sometimes needed to chain the trucks up to plough through seven or eight inches of hail. The Black Mesa is the place where I came the closest I have ever come to being fried by the elements.

Lightning is the lady in red for those of us who work outside. Beautiful but deadly. Students of Louis Battan know that lightning on the average kills more people than either hurricanes or tornados. They also know that the average bolt often exceeds a peak current of 20,000 amperes and sometimes exceeds as much as 100,000 amperes. The surge of energy can translate into temperatures as high as 30,000 degrees centigrade. I heard a story about a forester who has been struck by lightning six or seven times in his career and still managed to survive!

I've seen lightning sing through and bounce off the wires of a fence and walked through the splintered debris of trees that were in the wrong place at the wrong time and obliterated. Once, from the safety of a truck on Rampart Range Road near Colorado Springs, I actually saw a tree struck by lightning. The wood showered down over us a hundred feet away.

That afternoon on the Black Mesa I was stapling boundary signs up around a salvage sale of dead spruce. The storm was moving quickly, but I made a point of counting the seconds between lightning and thunder—five seconds to a mile. I figured to head to the truck before things got out of hand. The light filtered through the clouds in a way that initially turned the air into an amber-colored fog. A kind of snowy pellet then slanted in on a strong wind that came up out of the west. It wasn't snow but it wasn't hail either; it had the white color of snow and none of the icy hardness of hail. With the graupel in the air the light changed to a translucent, softly electric purple hue.

I still hadn't counted an interval that made me think the lightning was close-by when I looked down at the stapler and saw that little blue sparks were jumping around on it. The hair on my arms was standing straight up. I tossed the stapler to the ground and before I could drop all the way I was knocked over by the concussion from a shock wave of light. It was over that quick, but I laid there for a while.

There was no more lightning when I got up. I left the stapler and hustled back to the truck. I was safe by the time the real thrust of the squall line hit the mesa. The rogue bolt of lightning that bushwhacked me had come fifteen minutes before anything else hit. I sat in the truck for the next two and a half hours too scared to think, too scared to move, and too scared to drive back to town. The sky was sunny and clear by the time I headed home. On the way I considered the pros and cons of carrying a spare pair of shorts in my pack.

Along with the wet spring snows, early rains, thunderboomers, and autumn storms came the mud. Although a truck could get mired anytime, spring was the season we most feared. This was the tree planting season for the same reasons that it was the mud season. The idea was to get the seedlings into the ground while the melting snow and spring rains still maintained its moisture. During that wet and quirky season the Forest Service let planting contracts for hundreds of acres to independent contractors. Our job was to get the trees to them and inspect their work. The planting units were often located at the end of a pair of backcountry ruts that passed for a road in the dry season.

In any endeavor there are those who rise above the ranks of dilettante, journeyman, and even expert. Mastery must be the almost mythical mix of experience, natural ability, and flair. Mike McGuire was a master of making a truck go where he wanted it to go, and the unique blend of soils that turned the backroads of the Glade into a slimy, muddy skating rink each planting season were the test of his artistry. Above all other considerations was Mike's conviction that man was meant to drive to any place on this earth that he damn well pleased. To walk away from a truck hopelessly stuck in the ooze of some mudhole was a defeat for the human race, the way Mike saw it.

What made the Glade unique was the occurrence in the area of Mancos shale. When the rains came it turned to a grease that could be so slick that a truck parked on the slightest crown or smallest degree of camber might break loose and slide into the nearest ditch. Each of us dealt with the slime in our own way. The more timid chained all four wheels and slipped the transfer case into compound low, then crept cautiously along the slick tightrope of road. When the truck broke loose and slid into the ruts, as it almost always did, they'd sigh, get out, and go to work at getting the thing unstuck. The more adventurous added a bit of throttle to the process and generally made more headway.

Mike was a strong advocate of momentum. He scouted the road far ahead for any signs of trouble and the merest glint of sun off a distant mudhole was immediate cause to increase speed. He'd wrap it up as fast as he could go and scream into the mud, balls out. On impact, from the interior of the cab, the view turned to a very deep brown. Mud was over the entire windshield, mud was up over the cab, mud was flying twenty, thirty, forty feet in all directions. There was no concern for the machinery because at this point it became a vehicle for the human spirit, an extension of Mike's own body in a battle that would have but one victor. If he made it through, which he often did, he might turn to his seat belted, fear-stricken passenger and mention that, "Things had gotten a bit western, hadn't they?"

I was with Mike once when the mud won. We had just negotiated the Black Snag road, which in the springtime is not known for its ease of passage, and turned west toward some planting units. In a bottom just before the land began to rise there was a thirty-foot-long glide of mud. Mike hit it at about fifty miles per hour. The momentum carried us for about twenty feet before we came to an abrupt seat belt and harness snapping stop. It was abrupt enough that we sat there for a moment to regain the wind that had been knocked from our lungs.

Mike jumped from the cab first and surveyed the damage. A little crumple in the fender was the least of his worries. We had smashed into a set of ruts that were deep enough to high center the truck. The momentum of our run into the mud had been enough to bury both axles. The tires dangled helplessly four inches above the bottom of the rut. Mike cursed masterfully, then slopped through the mud to the back of the truck and threw the tools we would need for extrication up to the dry ground.

Among the shovels and picks was a tool from hell called a Handy Man jack. It's basically a stouter version of the bumper jack that used to be standard equipment in most American-built cars. Newer models, our truck included, had converted to the smaller, more handy hydraulic jacks that fit under an axle. These are fine for changing a tire on the Interstate, but for a truck already mired in the mud up to the axles they are worthless. This is why the Forest Service thoughtfully provided each truck with a Handy Man for the planting season.

The drill for getting a high centered truck unstuck is pretty standard. The wheels must be jacked up high enough so that rocks, branches, or anything solid can be placed under them and thus raise the axle off whatever it is high centered on. For a truck stuck on both axles this means jacking up each corner of the truck in succession to raise all four wheels. The rub comes with the Handy Man jack. First, rocks must be placed under its base so that it won't sink into the mud once the weight of the truck is on it. The actual raising of the truck is fraught with danger because the jack has the habit of flying out from under the weight if it hasn't been positioned just right or the rocks under it slip in the mud. People have been knocked unconscious and even killed by Handy Man jacks.

It took us the better part of an hour to get rocks under all the tires. When Mike finally slipped behind the steering wheel, punched it into reverse, and gave the engine gas, the carefully placed rocks flew out from under the wheels and the truck settled back onto the axles. We did it all over again and after another hour or so we had made about two of the twenty feet we needed to back out of the mud. We worked into the dusk and never gained more than five feet on the mud. Mike's shoulders slumped in defeat when he finally said we'd better just as well walk back to camp.

The next morning we commandeered a jeep with a winch and pulled the truck out. Mike hopped in it and rammed it into the mudhole again, only to get stuck again. I winched him out and he turned the truck around to go out the way we'd come in. He turned to me, eyes cast down in defeat, and we headed back to camp. He didn't have anything to say for the next three days.

I construe weather, in its broadest terms, to include the clouds of insects that storm over the high country of the San Juans. Their season is as predictable as the spring rains and autumn snow. I have learned the formula of the misery that their great swarms bring down on every mammal that is in the forest during their times. It is a formula with sequence and rhythm.

The flies come on quickly with the warmth of spring. The sheltered spots in the aspen and spruce might still have deep snowdrifts when the no-see-ums come into the air. These are the miniscule gnat-like flies. They are the black dots with tiny white wings that go unnoticed until their bites raise small red bumps on the skin.

It is possible that a few no-see-ums will still be in the air later, when the mosquitoes rise up from the moisture below the aspen. Their entry into the world is cause for concern, but the swarms seem short-lived. I am one of the lucky ones who does not react to a mosquito's bite. An average day in the forest during their time might produce seventy or eighty bites, but of those I might only get an itch out of four or five. This mountain variety mosquito seems strangely less venomous than those encountered lower down near town, which appear to have evolved to cause a special misery to humans. The high, whiny song of the mosquito around the ears is what drives me nuts.

As the mosquitoes begin to taper off, their flights are often mixed with the darting, deltoid shapes of the deer-flies and horseflies. These are the Darth Vaders of the fly world. The deerfly's eyes alternate stripes of orange and black that glow with a sinister fluorescence. Their bite is painful, but the Achilles heel of the deerfly is its lack of quickness. Their careful search for the perfect spot to drill their victims makes them an easy target for a punitive raid of slaps and swats. Were this not true, their time in the air would be a nightmare. The horsefly is the big brother of the deerfly but with less offensive glowing green eyes. Their bite is a bitch, but they seldom get that far and are known more for their ability to fill the cab of a truck where the window has carelessly been left open.

The deerflies mark the end of the warmup and things turn wretched in earnest with the advance of a small, brown fly with a pointy butt. It has been called a pecker fly for its affinity for certain parts of a bull's anatomy. The plague of pecker flies is not known so much for their bite but more for the sheer numbers of the flies themselves. They come by the millions. The harassment is due to numbers and their habit of burrowing into the hair of mammals. I have seen them so thick that they walked on the inside of my glasses and a breath that didn't include one in the nostrils or mouth was unusual. They drive the deer and elk to the point of jumping crazily through the meadows to rid themselves of the horde.

Along with the deerflies, and finally, come the black flies. These are darker and rounder than the pecker fly and resemble a somewhat smaller version of a regulation housefly. If there can be a greater number of anything other than pecker flies, then that number is black flies. Their harassment is unrelenting until the frost kills them off in late August.

The worst fly season that I have endured was when Randy Houtz, Curt Pfieffer, and I marked the Roaring Ridge Timber Sale off Roaring Forks Creek. It had been a wet season that year and the black flies came off early and in vast numbers. It was our habit to start work just after sunup when the cool high country air put a damper on them. That was the couple of hours that we figured to get some peace.

The onslaught was always the same. When the sun began to hit the forest around nine in the morning we heard a distant hum that seemed to crescendo louder and louder from under our feet. As it progressed we saw the flies coming out onto every twig, every leaf, and every blade of grass on the forest floor. We never ascertained the true origin of that great humming but took it to be the flies, all in unison, drying the dew and warming the cold from their wings. When all at once it ended we knew they were in the air.

We pulled our chins in like we were pushing into a gale and worked in a kind of irritable, claustrophobic prison that the flies created. We never stopped moving. At lunchtime we walked in tight circles, eating our sandwiches, and then moved on. By early afternoon we snapped at each other for no reason at all other than the cloud of buzzing flies. Flies covered our shirts, our pants, and our faces. A simple swat to the thigh was enough to kill twenty or thirty of the beasts, but that vacuum quickly filled with thirty or forty more flies. The only relief came with the coolness of the evening, the screens on our trailers at camp, and the whiskey we poured as soon as we got there.

Autumn brings the best in weather to the Rockies. The late August or early September frosts in the high country kill the flies with the possible exception of the yellowjackets, which still busy themselves drilling into the ground for winter. There is a change to the feel of the air almost as if it has lightened in anticipation of the load of snow it will carry later. This lightness clarifies things for that brief season, distances seem shortened, and the sun itself seems to radiate with more warmth.

I look for the snow to come in the late fall and take it as the signal that I will be laid off from work. We flounder around in the first few storms, put chains on, and slide on icy roads toward a few final errands, but the snow is absolute in the way that it closes us out of the mountains. Last year I was on the tail end of a timber survey in the aspen when the storm that would end my season came. I worked for a couple of more days in the windless silence of the falling snow that pulled me inside of myself the way that winter should. One afternoon I built a fire and sat by its warmth until quitting time while the quiet of that place grew. The leaves had fallen, the elk had moved to lower elevations, and it was still in the forest. The weather had declared the country closed.

NIGHTJARS

Disappointment Valley lies just to the north of the flattopped pine country called the Glade. It's a long slide down draws like Wolf Den, White Sands, Ryman, and Box into that brushy place. The pine gives way to Gambel oak, antelope brush, sage, and a host of "invader species" that point to overgrazing. A good bit of it is Bureau of Land Management land and supports the old story that in the beginning the Forest Service took the "good" productive land, meaning timber, and what was left eventually ended up with the BLM. The ranchers would disagree and so would the elk that come out of the San Juan Mountains to winter in the valley by the thousands. The elk were so thick that one winter the Division of Wildlife held a special December hunt to slick some of them off so there would be forage for the cattle come spring. It's called bringing the herd under management.

The valley is imbued with the kind of faraway, edgy

beauty of a place that is seldom scrutinized. The view from the Benchmark Fire Lookout up on Glade Mountain is magnificent. To the west and northwest the valley stretches broadly out toward Slickrock and the Big Gypsum Valley and farther still the Abajos and Canyonlands. To the east the valley wanders toward its origin through a roughed-up landscape of gravelly buttes and mesas. It tightens and sways through thicker and thicker oak brush until there isn't much left near the Buckhorn Lodge, which is one of few signs of human life and the location of that country's single radio telephone to the outside world.

The distant view looking that way is dominated by Lone Cone, a twelve-and-a-half-thousand-foot, perfectly conical peak that stands off alone on the divide between the San Juan and Uncompahgre forests. Beyond that are the rugged peaks of the San Miguels—Dunn Peak, Middle Peak, Dolores Peak, and three of Colorado's fourteen-thousand-footers—Mount Wilson, Wilson Peak, and El Diente. The big, flat tabletop known as Black Mesa finishes the scene by hammering at the southern edges of the range.

That is the overview, but the mysteries confine themselves to the smaller draws, benches, and bends that find their way into Disappointment Valley. It's a place that has been peopled in a stingy way, mostly by ranchers, but has seen its share of desperados and the likes of those who don't fit smoothly into the machinery of town life. The chances they've taken are little more than a whisper in a big place and show up in the form of barbed wire, abandoned mining claims, broken liquor bottles, and the memory of water passing under the Daddy Williams bridge. Those who made it may have passed down a homestead or a range allotment or maybe a section or two of scrub along Disappointment Creek. The survivors have gotten bigger, needing more land for more cattle, and the others are just gone.

I took to going down to the edges of Disappointment Valley the summer I worked out of the Glade Guard Station. It was a rocky drive down the Black Snag road with no set boundaries. I figured the valley began where the pines left off, but there was no line of demarcation—simply a change in mood when the forest thinned to a certain point. There was a place where individual trees stood far apart from each other. They were the few that ran up against whatever barrier it was that held their kind out of Disappointment Valley.

I am a ridge person by nature. The idea of running a high line and gathering in the big picture suits me, but I have learned to play a smaller, more repetitive game. I make a point of seeing the same thing as many times as I can stomach it because I know that connections are being made. Look at the same species of lichen a hundred times and you will know some things that are unexpected.

There was a change in perspective in the valley that was as simple as the difference between two birds. Up on Glade Mountain we watched nighthawks carve up the evening sky in huge buzzing arcs. Disappointment Valley was poorwill country. Both birds are from a family known as nightjars or goatsuckers and as a group are known more for their nocturnal calls than anything else. Their lives are a mystery of the night and that suits the land around Disappointment Valley.

The first nightjars I became familiar with were the whip-poor-wills that are a sort of eastern equivalent of the poorwill. The nightjar family is one of a few where many of the species are named for the call they make. I associated the birds with the humid evenings in the woodlands of Virginia. There were spots where I could depend on hearing the loud WHIP-poor-WILL or sharp whip-whip-whip stab through the darkness. The calls put meaning into the name nightjar, which breaks down to night + jar, meaning roughly to jar the night or simply make a harsh noise in the night. The family's other nickname, goatsucker, is come by with more difficulty but relates to a belief, possibly from Europe, that members of this family suck on the teats of goats after nightfall. This hasn't been proven true, but the sight of a whip-poor-will's huge gaping mouth, used to capture insects on the wing, gives credence to the idea. At least it is *big* enough.

An entire lifetime of humid nights can pass in Virginia with the voice of the whip-poor-will as the only tangible evidence of its existence. I chose to try and see one early on and accomplished it on a Boy Scout campout in the Appalachian Mountains. It was twilight fading rapidly to darkness when I heard the voice close-by and took off after it. I was crawling through a mess of honeysuckle and came up for air in an opening and face to face with the bird. It hadn't left its roost yet and was sitting lengthwise on a limb, which I came to find out later is characteristic of the nightjar family. Even close up the bird was hard to see, being perfectly camouflaged. The whip-poor-will called out a few more times as I stood motionless, then took silently to the air. Their wings, like owls, are muffled by soft feathers to avoid scaring off prey.

I got to where I could occasionally locate whip-poor-wills during the day when they slept, eyes clamped shut, either on the ground or more commonly on a low horizontal branch. The bird is so tough to see with its cryptic markings that I often came close to stepping on them. In the east the bird is still sometimes associated with the larger and less secretive nighthawk, which was actually thought to be the bird making the calls until the early 1800s.

The other great nightjar of the east is the chuck-wills-widow. This bird has the same feel as the whip-poor-will—the soft, hard-to-see plumage, huge mouth, and a name that sounds like the call. The difference is that the chuck-wills-widow is a much bigger bird. The chuck-wills-widow tends to be more of a southerner than the whip-poor-will and my home in northern Virginia was near the end of its range. The long CHUCK-wills-WID-ow even seemed more southern, almost with a drawl, and the sound of it was every mystery I had ever reckoned to come with the night.

There's an interesting hitch to the chuck-wills-widow's diet. While it commonly tends toward the standard night-

jar fare of moths, beetles, and other flying insects, it sometimes gobbles down small birds, with warblers seeming a
particular favorite. Chuck-wills-widows were not common
where I lived and I have never seen one. I know them by
the call and an odd association in my mind with the Con-
federacy. They conjure up the name of John Singleton
Mosby, the Confederate raider who stalked the woods
around where I grew up and was called the Grey Ghost.

I am caught by the nightjars and it did not leave me
when I came west. The nighthawks at the Glade were
hard to miss. Sometimes they flew during the day and their
pointy wings with the wide white wing patch were easy to
spot. More commonly they came out at dusk, often over a
point or ridge, and hunted. There are a lot of them out
there and the cowboys call them bullbats after the booming
roar that is made by the wing feathers during particularly
steep dives, most often during the breeding season. It was
the poorwills that I was unprepared for.

Of the common North American nightjars the poorwills
are the most mysterious and least heard about. It could be
due to the fact that they are a western bird common to
the arid, brushy uplands. Places like Disappointment Valley.
Places where there aren't many people to hear the pooooor-
WILL that comes with the twilight but ends by full darkness. Places where people don't talk so much, especially
about noises on the edge of darkness. I picked up a
standard general duty dictionary once and looked up my
favorite nightjars. The whip-poor-wills were there, chuck-
wills-widow had an entry, nighthawks were represented,
and even the more general terms nightjar and goatsucker
appeared. There was nothing for poorwill. The dictionary
simply went from "poor white" to "pop" with nothing
in between.

My first trip to Disappointment Valley was full of
poorwills. I'd left the guard station after supper and made
the valley in about thirty minutes. The point of the trip
had been to just take a look around, but that wasn't easy.
The edge of the valley was grown thick in oak brush and

I couldn't find a knoll or a hill to scramble up on for a bigger view. There's a precision of pattern in the way oak grows that can almost always be traced back to the amount of water available to the plants. There in Disappointment Valley the clumps of oak were evenly spaced fifteen to twenty feet apart. It was the kind of exquisite cover that the big mule deer, particularly the bucks, seek out. The ins and outs of that kind of maze leave plenty of escape routes and ample forage. I've seen bucks in those places that I mistook for elk, but hunting them requires a patience that verges on trance and that is why they are there.

I never found a good spot to be and finally lowered the tailgate on the truck and waited for the twilight. It came gradually that clear evening almost like a tide would come over the beach taking a little more sand with each wave. A point came where things were indistinct and I heard the first poorwill of my life. I'd never made an exacting hobby of nightjars and at the time didn't know anything about poorwills, but the almost sad call of the bird reminded me of other twilights spent listening the night in. I strained to locate the bird but couldn't. Within minutes I heard dozens more poorwills echoing one another and finally the air was saturated with their sound. The birds must have still been on the roosts, under cover, because I walked back and forth on the dirt road trying to see just one of them, but they remained invisible. It went on until it was dark enough to see the stars, then the chorus stopped almost completely on cue.

I had figured that the call must be some sort of night-jar if only because it resembled the whip-poor-wills I had heard so many times in the East. I even thought it might be some western relative that had clipped short its speech to go with the sparseness of the land in the same way that the old hardscrabble ranchers spoke in fewer and fewer sylla-bles each season. Whatever the reason, the twilight had been full of the bird's calls—hundreds of them it seemed like.

I turned the truck around and headed out. In the headlights I spotted the poorwills. At first I noticed just their eyeshine, which reflected a deep, almost pagan pink. I'd seen the deep green reflections off a deer's eyes before and even the amber of a cat, but these were new colors and they were everywhere. I picked up the fluttering birds next. They jumped up from the road in two- or three-foot arcs like huge renditions of the moths that they were chasing. I left the lights on and got out of the truck. The poorwills let me get close enough to see that they looked very much like the whip-poor-wills I'd known, only smaller and with maybe a little grayer plumage. They flew silently bending the air with their short tails.

I spent as many twilights as I could in Disappointment Valley and the poorwills came out faithfully with each one. I even managed to sneak down once during the day in an attempt to kick up a few of them but couldn't find any. It could be that they were there, just under my feet, but that their plumage camouflaged them so well that all they needed to do was sit tight and they would become just another rock resting on the ground. For all I knew maybe they *turned into* rocks during the day.

Over the winter the poorwills stayed with me, more like a fragrance that floated through my mind than any kind of exact image. I decided that I would go back to graduate school in ornithology and write my thesis on them. It looked like the kind of scientific deadwater I enjoyed. I searched the literature and found that only one fact had briefly rattled the poorwill's cage of obscurity. It appeared that a couple of the birds had been found during the winter in a rock crevice out in the California desert and that they had been sleeping—deep sleeping. The fact was they were in hibernation. Hibernation, up to that point, had been unknown in the bird world. There were cases of torpor, a reduced state of metabolism known to occur in some species, most notably hummingbirds, but this lasted at most for a night's time or a few days of unseasonable cold. The poorwills were sleeping the better part of the winter.

Poorwill migration had been sketchily documented, but the two birds in California presented a problem. It might be that all the birds were *not* wintering in Mexico. Biologists went into the field and measured the sleeping birds' body temperatures and found them to be very close to the surrounding environment—like hibernating lizards or snakes or frogs. This was good stuff. The kind of thing that gets a young biologist out of the assistant professor ranks and up to associate professor and tenure country. They started grabbing poorwills and bringing them into the lab and stuffing them into refrigerators to see if they would hibernate. They wrote papers. The birds, indeed, did seem to sleep *very* heavily in the winter. Of course, they could have just asked the Hopi Indians, whose name for the poorwill is Hölchko, which roughly translates to "the sleeping one." Some bird books now state the migratory status of the poorwill as unknown.

I had a meeting in Flagstaff with an ornithologist at Northern Arizona University who was interested in sponsoring my graduate work. He hadn't heard my ideas yet on poorwills and when I told him he said there was no way. "We don't know enough about them for you to study them. It would be too difficult," he said.

It was an interesting problem. Apparently the development of a statistical model would be dicey and that was the new buzz word in biology. I asked why couldn't I just go out and *watch* the birds, maybe laying a foundation for future study. That wouldn't do because the world needed numbers, not old-timey naturalists. We needed to quantify, quantify, quantify. I gave up on graduate school but kept my eyes open.

There is a wonderful series of ornithological papers on the life histories of North American birds that was collected by Arthur Cleveland Bent. The bulletins span decades in the early and mid-twentieth century. The thirteenth in the series is *Life Histories of North American Cuckoos, Goatsuckers, Hummingbirds and Their Allies.* There are some statistics included in the monographs, like numbers and sizes of eggs in an average nest, but more importantly the papers are full of observation. The great ornithologist Elliot Coues contributed this on poorwills in 1874, "This cry is very lugubrious, and in places where the birds are numerous the wailing chorus is enough to excite vague apprehensions on the part of the lonely traveller, as he lies down to rest by his campfire, or to break his sleep with fitful dreams, in which lost spirits appear to bemoan their fate and implore his intercession." A Mrs. Bailey in 1928 described the poorwill's call, ". . . like the delicious aromatic smell of sagebrush clings long to the memory of the lover of the west."

There is a likeness to poorwills in the drift of certain men into the forest. They are the kind that have somehow managed to end up on the edge of things and are carried into the remote places for vague reasons. They are a mish-mash of ramblers, hermits, ne'er-do-wells, sociopaths, half-crazed and crazy souls. They are not truth seekers in any conventional sense and you won't run across any autobiographies of their visions. These are journeys of relief and for damage control. A search for balance. And their legends are like the echo of a poorwill's voice up against the night. They are wildmen.

The idea of wildmen and the forest is nothing new. It's been said that during the Middle Ages in Europe lunatics and the occasional crazies were simply led a safe distance out of town, into the woods, and let go to fend for themselves. There is an entire literature of wildmen as old as the Greek myths or woven smoothly into Buddhist texts. They are here, real or imaginary, hermits like Hanshan and Shih-te, Don Quixote, Rousseau's Noble Savage, Robinson Crusoe, Tarzan, Sasquatch, the Abominable Snowman, Jesse James, Jeremiah Johnson. . . some with hair-covered faces, some kind, some wise, some violent, some raised by wolves. . . all riding the tides of our collective thoughts about freedom, passion, and sanity.

For every international wildman there are hundreds more local or regional everyday wildmen. Legends that almost always grow out of facts. Men that run wild in some isolated valley, then disappear or are caught but grow in stature to the point that every man, woman, or child in that country has seen him or talked or shook hands with him. Their children and grandchildren remember. Word gets around. Here in the Rockies and Basin country we take our wildmen seriously; it's open country.

They say Navajo Sam got his name when he was living down south. That was before he took up living in a little make-shift outfit on the Dolores district of the San Juan National Forest. No matter where the name came from it fit perfectly that first day when he came out of the brush on the Navajo Lake Trail. The trail was the main thoroughfare for backpackers heading to a high country lake of the same name that was nestled in a basin surrounded by El Diente, Mt. Wilson, Gladstone Peak, and Wilson Peak up in the San Miguels. It was the summer of 1982 and the year that Navajo Sam decided to start holding up hikers for their lunches. He had a gun, which also fit perfectly; here in the West we also prefer our wildmen to be armed. It emphasizes the point of who is really in control.

Navajo Sam's first victims were a couple of doctors from Grand Junction, Colorado. They said that he never really pointed the gun at them but instead laid it across a rock. They chose to give him their tunafish sandwiches without a struggle and listened to him rant and rave about big government, big business, and the rich guys getting richer. The Forest Service got on the case as soon as the word swept down the Dolores River Valley and throughout southwest Colorado. The newspapers picked it up, too, and it went out on the wires. It was absolutely irresistible.

"Navajo Sam stakes out his claim to the Navajo Basin." If not in the open, then secretly we all kind of applauded him. I don't believe it was the violence that turned us; people stick guns in other people's faces every day and you can get all of that you want by simply watching the Albuquerque news. It had to with the idea of one man making a stand against overwhelming odds from a wild and remote hideout. Navajo Sam was fighting economics, the government, society, and pretty serious

depression. And he was in a position to elude capture—
that's the other thing we want in our wildmen. They should
be woodswise and crafty, to a superhuman level if possible.

It went on for several months until the hunting
seasons came around. Forest Service special agents figured
that while Navajo Sam was somewhat of a danger, the
greater threat might be that some half-crazed hunters would
ride into Navajo Basin, grease him, then pack the body
out as a public service, hoping to collect some kind of
reward or at least become part of the legend. The agents
disguised themselves as elk hunters and picked him up
near Wood's Lake.

They threw him in jail in Cortez and charged him
with aggravated robbery and felony menacing. A grass roots
movement developed to free Navajo Sam and bumper
stickers began to appear that said "Free Navajo Sam." A
bank account was started to help in his defense. Three
weeks later a Dolores County judge freed Navajo Sam for
lack of evidence. He was a folk hero. A wildman.

In their introduction to *The Wild Man Within*, a
collection of essays on the image of the wildman from the
Renaissance to romanticism, Edward Dudley and Maxi-
millian Novak described the wildman as "belonging to the
region of the mind that treasured freedom over control,
nature over art, and passion over abstract reason." Navajo
Sam had walked boldly into that wild country.

It didn't end with the cheering in that Dolores County
courtroom. Another warrant was issued when charges
were refiled a few weeks later. Navajo Sam high-tailed it
out of the area. Even today you'll hear the stories that
he's still up there, though, off living in some remote, wild
valley, where we'd want him to be. Actually he moved
to Wisconsin.

I was working on the Dolores district during the
Navajo Sam summer and when the case was closed that
autumn I went over to the Hollywood Bar with one of the
Forest Service special agents. He was glad Navajo Sam
had taken off.

"The hell with justice, we didn't want to be the bad guys that brought a legend in. It's rotten PR and that could hurt us more than Navajo Sam ever could," he said.

The ridgerunner had been gone for the better part of fifteen years when I moved to northern Idaho in 1980, but we hadn't even unpacked our goods in the house we rented in Troy before I heard the legend. My new neighbor, an ex-logger, dropped by to see if he could help out. In the course of conversation he glanced over the rolling hills and mentioned matter of factly that he'd had a cousin who'd seen the ridgerunner over in the Clearwater country. I asked him who the hell the ridgerunner was and he looked at me like I'd just landed from Venus.

The ridgerunner is as much a part of that thickly timbered country as the small logging towns that cling to its edges — places like Bovil, Clarkia, Pierce, Weippe, Headquarters, and Avery. If the ridgerunner isn't part of the conversation on any given day be assured that he is stalking the edge of people's consciousness like the fog and the rain and the forest.

William Clyde Moreland came into the Sawtooth Range of southern Idaho in 1932. Up until then he'd lived a hardscrabble life going from reform school to reform school and sometimes spending a little time in jail for burglaries and the like. He'd managed a fifth grade education somewhere along the way. Odds are he came into the country as a last resort and that first year wintered over in the Chamberlain Basin. There is hard evidence that he tried to steal an airplane down there, but when he jumped into it and started turning keys and cranking switches he couldn't get it to run. He didn't know how to fly anyway. He spent the winter living off deer that he snared using telephone wire he'd ripped off from the Forest Service. He stayed in the backcountry from then on, slowly moving north into wilder and wilder places.

He was seen on the upper reaches of the Selway River in 1936. At some point he made an impression of a Forest Service key in a bar of soap and filed a key out of tobacco tins. He used the key to enter the Forest Service cabins that were located along the trails throughout what was then pretty much roadless backcountry. They say he usually left the cabins a bit messy and had a real taste for jam. He seldom slept in them but rather grabbed some stores and took off back into the forest. Depending on who you talk to, the ridgerunner is regarded as a gentleman who "borrowed" what he needed to get through hard seasons outback or he was simply a wacko thief.

Among the woodsmen, the ridgerunner developed a reputation for cunning, stealth, and endurance under unimaginable hardship. There are stories of his snowshoe tracks disappearing into thin air, his boot prints turning mysteriously into a gaggle of elk tracks, and long winter treks sometimes covering hundreds of miles. There is some truth to all of it. He *was* really out there.

From 1937 to 1942 Moreland travelled the backcountry of the Clearwater and St. Joe national forests, spending a good deal of time around the old Roundtop Ranger Station. By this time the Forest Service knew that somebody was entering their cabins, but it wasn't until the rangers got together that they came to the conclusion that it was in fact the same man. A theft in 1942 of some food, clothes, and bedding at a trail camp led the Forest Service to believe that the ridgerunner was Charles "Baldy" Webber, who was wanted for attempted murder and considered dangerous. The Forest Service set out after the ridgerunner. In 1944 two Forest Service workers ran into Moreland at the Flat Creek cabin. He was cooking dinner and invited them to chow down with him. Afterwards he left and the Forest Service guys called the boss and asked him what to do.

The Forest Service got hold of Morton Roark and Mickey Durant, both skilled Service woodsmen, and asked them to pose as trappers and roam the country in an attempt to bring Moreland in, who they still thought was Webber. They tracked and trailed him into the spring. There were a few narrow escapes. Finally, they tried again in the winter of 1945 and captured him near the Skull

Creek cabin. Instead of the highly dangerous desperado they expected to find, they found the ridgerunner, living under a piece of canvas in the middle of winter with a wet sleeping bag. He'd lost most of his teeth and had a touch of scurvy. He'd been running the ridges for thirteen years.

He ended up in jail after a short visit to the state mental hospital, where he was declared a bit on the antisocial side but not really very dangerous. The folks loved him. Anyone who could make it for thirteen years in northern Idaho, summer and winter, was okay with them. He explained the disappearing snowshoe tracks very simply—he had a harness on both ends of his snowshoes and simply turned around and walked back in his tracks. He took to the streams now and then to avoid detection and always travelled on the ridges up above and parallel to the trails. The boot prints disappearing into elk tracks? Be it legend or not it is said that he had a short little pair of stilts with elk feet on the bottom.

After he got out of jail he managed a job with the Potlatch lumbering outfit on the Camp T Flume but blew up a tractor with dynamite when the boss got on his nerves. A local jury wouldn't convict him. In 1952 he took a few shots at a Potlatch foreman and ended up in jail for six months. When he got out he took to living in the bush again and stealing from logging camps and the Forest Service. He also started writing weird letters to the governor and regional forester. In 1956 he was caught in the Canyon Ranger Station that was unoccupied at the time. He ended up in jail for six months again and headed right back into the mountains when he got out.

He eventually stole a .45 pistol and some goods from the Skull Creek cabin in 1957. The FBI was called in and the ridgerunner ended up in the loony bin in Orofino. He escaped in 1959 and was returned. He was released in 1961. A ranger came across him wandering the backcountry after that and asked him what he was doing. He said he wasn't going to stay but had just come back to see if he'd been dreaming all those years. They say he died in 1964, but nobody knows for certain.

I've taken a good bit of the ridgerunner story from an account written by Ralph Space in *The Clearwater Story*. He was the supervisor of the forest from 1954 to 1963 and in on much of what happened to the ridgerunner in his final years. There are other renditions, like the one written by Bert Russell in *Calked Boots*. The facts don't differ radically between the two accounts, but they are worlds apart. Russell, who's logged, cruised timber, and worked the mills, sees the ridgerunner as basically a good ole boy who managed to slide by the government for years by outsmarting them and one upping everyone with a superior kind of woodsmanship. Space sees the ridgerunner as insane and because of that enduring untold hardship. I wrote him in 1982 and asked if he knew any more about the ridgerunner. In his reply he said he had nothing to add to the story but warned me that most people had made Moreland a hero, "describing him as a poor but clever woodsman who eluded all efforts of the mighty Forest Service to catch him. Such was not the case. . . . I hope you do not try to make him a hero," Space said.

That would be the obsession. The voice of reason in one ear, the wildman in the other. There are no real people in any of this. It is the kind of thing that floats around on the wind. The grand legend that has been known to outlive governments.

More recently it has gotten ugly. Claude Dallas killed two Idaho Fish and Game Enforcement officers on January 5, 1981, and busted off into the wild country. Some said that he was so strong an outdoorsman that he could live indefinitely out there. The stories started. The legends formed. The wildman crawled out of his box. In the summer of 1984 Don Nichols and his son Dan kidnapped Kari Swenson, an athlete from Bozeman, Montana, and headed into the mountains. They killed a man in the process and wounded Swenson. In a news interview Don Nichols's sister Betty said, "Don's loved the mountains all his life and he knows how to live in them better than anybody I ever knew. They won't starve him out. He doesn't smoke and he doesn't drink, so he doesn't have to come down for supplies. He says he's always warm up there, even in the winter." The wildman scratching at the door.

All of this is unexplainable. These *people* must be crazy. It's the stories and legends that live a life of their own. The idea that there is something like absolute freedom and that it resides in a wild place. A thread so strong that utter deprivation seems acceptable and even murder can get by. The wildman is the story that lives on that edge between twilight and darkness. The shadow of a poorwill jumping through the headlights. Something we don't know enough about to study. A mystery recognized by its voice in the night. A nightjar.

MEDICINE

In northern Idaho being lost was not totally unusual. The timber was thick and ran clear to the top of the mountains. It wasn't like other places in the Rockies where there's a bare, stony ridge cropping out here and there that can serve as a point of reference. On any other day being lost would have been fine. We'd have made a point to not leave the drainage that we were in when we discovered our predicament and slowly have retraced our paths until we came across something familiar. If nightfall got in the way we would have simply built a fire and laid around until daybreak. Odds were they wouldn't have even missed us back at the Musselshell Work Center, at least, for the first night. There would have been no panic. No telltale quickening of pace. No cussing. No swearing. No whining. As they sometimes say in the Forest Service, "The pay's the same."

But this was no regular weekday. It was Friday and

121

the crew was moody, on the verge of hostile. In that great tradition of people who work in faraway, even lonely places, their minds had been set on blowing out of the woods at full speed and landing in the nearest town and getting drunk on their asses.

As the crew boss it was my obligation to help them fulfill the tradition. I could see the panic just beginning. We retraced our steps to a huge western cedar that had fallen over a narrow creek. This had been the reference point we all agreed to, but the rest of the country made no sense. We started working wider and wider circles around it and marking our way with surveyor's flagging. I overheard some of them talking quietly about taking off cross country on their own to follow a straight, due north bearing until they hit the road six miles north. From there they would hitch into town. Six miles in that cut-up and brushy mess of country would have taken them all night and some of the next morning to cover. I counciled prudence.

Finally we came upon it. The creek with the big cedar had forked above us. On the other fork, in a roughly identical position, was another large fallen cedar across the creek. This was our real reference point, the one we had carefully marked in our minds that morning when we took off. It was all a twist of fate. A simple mistake. We took off on a beeline to the truck and made it out of the country before dark, the only damage being that the crew tore the town apart a couple of hours late.

With the possible exception of tearing small towns apart, the idea that a celebration of some sort is in order upon leaving wild places is not new. In the very old days it had more to do with feasts, storytelling, and probably some serious lovemaking or at the very least attempts at serious lovemaking. It also usually involved some attempt at consciousness alteration. Maybe ten or twenty drinks or a friendly hallucinogen. At least a strong cup of java.

It's a squirrelly equation, nowadays. Here we bust our asses off to get out into the backcountry and make a stab at a little serenity, a little calmness in the face of adversity. Maybe sign on for a season with the Forest Service just to be outside, under the big sky, and the first thing we do after a week in the woods is come into town and get drunk out of our minds.

I thought for a while that the obligatory "back in town" drunk served a kind of reentry into civilization purpose. I'd idealized the being in the wilderness business to the point that I figured that we got drunk to get resynchronized with the regular world. Going down to the bar and having a few while you watched the evening news accomplished that. It could be that a certain dulling of the senses was necessary for survival in town.

The flip side has a more serious, even darker, tinge to it, at least if you place yourself in that group of folks who reckon that being out and away from it all is the only way to live. I'm one of the people in that group. It is possible that all the beauty, all the magnificence, all the grandeur, all the serenity gets a little boring and puts a bit of a celebration in order.

There really is nothing wrong with a little boredom. Some of my best times have occurred when I was bored. These are the times when I've studied, I mean really studied, how it is that a yellow-bellied sapsucker lines up each of the holes he pecks in a tree. Or gotten involved in ant ecology by slapping a fly out of the air, dropping it on the ground, and watching the ants come from nowhere and drag it off. Boredom is only ugly when it leads to despair and that only happens if you think about it. Otherwise, consider it enlightenment.

I've become a connoisseur of the small town watering holes. The diners and bars that stand alone by the highways or on the edges of the hamlets and towns that back right up against open country. I like them because anonymity is impossible. It may be that no words are ever spoken, but assessments are nevertheless made. I've had the same kind of feeling when I was alone and deep in the forest.

We stayed at the Rico Guard Station when we marked

the Morrison Creek timber sale. Actually, it had once been the ranger station in the days when Forest Service districts were smaller, more family-like affairs. We bunked in what had been the ranger's living quarters. It was a roomy, thick-beamed log cabin with a huge stone fireplace. A smaller district office and barn were located to the south of it.

The guard station was the epitome of what most Forest Service people still thought the Forest Service was—a log cabin in a deep valley on the fringe of a small mountain town, which in this case was Rico, Colorado. It's a fantasy that has long since given way to the bigger, higher powered district offices that seek to concentrate the expertise necessary for doing the paperwork required in a modern bureaucracy. Word is that there is still some resistance in the backwaters of Montana and possibly Idaho where a few of the district offices are reminiscent of the old days and, in fact, similar to small town watering holes.

The Rico Guard Station enamored us not only for its history and tradition but for its more practical aspects— like beds to sleep on and running water (hot and cold) and heat. After a summer of campfires, latrines, rain, and sheepherder's tents we were ready to come in out of the cold.

Most of all we liked being just two miles up the road from Rico and the Rico Hotel. In the heydey of mining in Rico the "hotel" had been a dormitory for miners. The rooms were tiny and no more than cribs that caught the dreams of the sleeping workers. After the collapse of the mining industry the building stood dormant for a number of years before it was renovated into a hotel. Walls were knocked out and the rooms enlarged and redecorated in a Victorian style. A combination restaurant/bar was added with a wood deck and geraniums. The idea behind it all was to catch the trade travelling Colorado State Highway 145 to and from the skiing in Telluride, or to and from Lizard Head Pass, or to and from anywhere. A local hangout was born.

We favored the hotel over its main competitor, the

MEDICINE

Galloping Goose, which had the tremendous advantage of a pool table, not so much for reasons of loyalty but simply because it was new and we were new to the country. It was struggling and we saw fit to drop in when we could and have a plate of the endless spaghetti special that simmered on the stove for weeks at a time and drink beer.

That late summer there was a young couple running the place for the owner. The man was a cowboy out of work and more inclined to roping than cooking. He figured to guide elk hunters in the fall. In the meantime, he smoked cigarettes and supplied local color. We did some serious drinking with him occasionally, but it was always unplanned.

There was an old guy who came in now and then and played ragtime on the piano. He'd lived in Rico most of his life and had played the joints when the town was booming. He played at the hotel for free when we knew him. The music came simply out of a love for times that were gone. There weren't even enough families living in Rico that year for them to open up the elementary school. He was playing the night it cut loose.

There is a powerful medicine that goes with any hot night down at the bar. It is no more explainable than a gathering of clouds that gives way to unpredicted weather. This night was no different. A couple of Kim's friends had shown up on Harleys from Wisconsin, which might have made a difference, but there was no forewarning of a celebration. The bikers were tired. The crew was tired. We were looking for a little chow and a few beers.

We had a few before the idea began to wash over us that it was not going to be an early night. No committees were formed, no consultations conducted, no policy enacted. We were half-sloshed before the piano player showed up.

There was still room for an organized, legal retreat to the safety of the guard station when the honky-tonk music started, but along with the music a road-weary couple from California walked in. When the lady started taking Polaroid snapshots of us all because we appeared to be

125

SEASONAL

authentic, our chances diminished. All hope faded when she ran excitedly from table to table with the photos that showed a strange light in the upper right hand corner. She figured the oddity to be the tracks of a ghost. Her husband winced and ordered a double.

We all broke into an extended chorus of "California Here I Come"—like CAL-LEEEE-FORNEEE-YAA Here WE Come. It echoed the pleasant west-slope-of-Colorado hostility that sometimes occurs in the face of California and ghosts. The old piano player picked up the tune. We were off, all of us, including the Californians, singing together. The lady from California continued campaigning, from table to table, for at least the *idea* of ghosts. And we started to believe.

The rest is a familiar history for connoisseurs of backwater bars. More beer, shots of Wild Turkey, shots of Dickel, men walking other men out to the pickup trucks to show off rifles and pistols, more shots, crazed dancing on the bar and tabletops. Through it all slid the idea, mind you, just simply the idea, of ghosts roaming the streets, buildings, and nearby abandoned mine shafts of Rico, Colorado.

It also became apparent that the woman from California was some kind of a singer, and the old piano player, who'd probably known some of the ghosts, also knew every tune she wanted to belt out. He gloried in it. It was one last chance to get at how things must have been.

At two in the morning the cowboy turned bartender came in with the owner and said that they had to close the place down by law but that he'd make a bit of an exception. We could stay there, drinking, if we'd agree to having the lights turned out. We lit a few candles.

It broke up at four or five in the morning. I'd like to say that we were too scared to drive and that we weaved our way back to the guard station on foot, but the fact is that we drove those deserted couple of miles in a careful frenzy. There was timber to mark in the morning.

There has been an unwritten law on every timber crew

126

I have ever worked on. We work hard and we party hard, but by golly we *always* make it to the job the next day. It's a matter of pride, whether you get anything done or not. We were up at seven the next morning.

Jim, a big strapping farmboy from Wisconsin who we called Jimmer, usually drove the truck, which was a standard pickup with a crew cab that seated six. We called it the six-pack. He walked over to me that morning and gave me the keys.

"You drive," he said, then puked.

There was never any talk of ghosts when we showed up at Jim Wagoner's place on Groundhog Reservoir. Mostly we talked about his biscuits. Jim made the best biscuits in all of the Groundhog Mountain and Black Mesa country.

We'd met Jim the first year I worked on the Dolores district timber crew and were marking the Little Fish aspen sale out on Groundhog Point. The district had contracted with him to let us park our trailers at the little fisherman's campground and concession that he ran on the south side of the reservoir. Along with that he also operated the gates on the dam and opened and closed them when he got the word from the Montezuma County irrigators out in the beanfield country to the southwest.

We were setting the trailers up when we heard this tremendous blast from over near Jim's cabin. At first, we thought one of the propane tanks on the trailers had blown up. That was until we spotted Jim in a cloud of smoke on his back porch with a .54 caliber Hawken muzzleloader in his hand. He smiled and waved.

"Just trying it out," he yelled.

Jim was living in the cabin with another out-of-work cowboy who was driving in a log truck at the time. The two of them made sure that we all knew we were welcome over there anytime. In fact, Jim suggested we show up for dinner in a couple of days. We did and that was the first taste we had of the legendary biscuits alongside of a great elk stew. Jim could cook.

"It's all in the lard. You gotta use lard in biscuits," he said.

Jim could also play the guitar. He'd been in a pretty high powered country-folk sort of band in Oklahoma until the pressure had gotten to him and he took off for wilder places. He'd ended up relatively broke and faraway on the shores of Groundhog Reservoir, at least, in the warm months. He wintered down in Dolores. Now and then he travelled around and played in some of the smaller mountain towns.

We got in the habit of showing up at the cabin and drinking whiskey and playing and singing songs. Jim could just pull a tune out of the wind and make up lyrics as he went. He'd sing songs about everybody on the crew and they were always funny. I don't remember any totally crazed times at Jim's, which could be good or bad, depending on how you look at it. What I do remember is getting the feeling that things were somehow like they were supposed to be whenever I ended up at his place. He had a way of winding his work, his music, his kids, and his life all into one big giant ball that reminded us of some of the dreams that we'd had when the word was going around that the best way to live was out on the land. I still run by his place for a snort or two every time I'm in that country.

The time always came when I'd end up coming into the towns and the cities, sometimes quicker than I wanted. The traffic and the lights were the first shock. I occasionally took to hanging around the country-western bars in places like Durango, Colorado; Pierce, Idaho; Orofino, Idaho; Dolores, Colorado; and Salida, Colorado. Places like the Hollywood, the Sundowner, the Strater, the Club Reo. I mostly liked the off-times. Two-thirty or three in the afternoon was all right.

The practical reasons for being in a country-western bar in the mid-afternoon can be considered questionable. Sometimes it was as simple as nothing better to do. But there are other reasons. There is a certain dreamy clientele that shows up then that I am drawn to. They don't tend to be the hornier crowd that comes later looking to "meet someone" or the "I just got off of work and I need a belt bad" group. It's more likely that the afternoon crowd will be the drifters. Maybe a gold prospector in town for supplies or an outfitter moving horses to another camp or a novelist.

It's best when the group is small, say no more than four or five, and gathered in a loose knot around the bar like so many rustlers backed up against a campfire for warmth. The talk tends to be measured and in the curt whispers common to the kind of people who are alone enough that they aren't always sure that the sound coming out of their mouths is their own. Much of the time it is conversation about the simple facts of life.

"I winched some slick out of the mud up in Horse Gulch, you know, near the bend at the lower meadow."

"Damn dog. I liked him until he got mean. Ended up giving him to a lion hunter."

"I've shoed good horses and bad horses and I'm too damned old for anymore bad ones."

"I thought it was in the carburetor, but it's still doing it."

"That was before my wife left me, so it was awhile back."

A lot of the time the sentences just trail off. Statements thrown into the air that aren't intended to be answered or even evaluated. Just something in the wind. And occasionally there is the sad story that is little more than the marking of time in relation to which husband or which wife or girlfriend or boyfriend it was then.

The knot unravels about the time the bar begins to fill, say five o'clock.

The other side of the coin would be Saturday night at a place like the Sundowner in Durango. If it was good the parking lot was full of pickups and the band was hot. It drew people out of both the country and the town, mostly just for the two-stepping or the waltzes or the fast stuff. They dressed up. The older couples waited for the slower buckle polisher tunes, then eased out onto the floor. There was something to seeing a forty- or fifty-year-old companionship on the dance floor.

The younger men and women worked the floor with the lukewarm method common to searchers. They scanned the bar and tables like smokechasers who watch the woods day in, day out for the telltale signs of anything that could mean heat. Now and then they found it.

Like any big dance there was always the possibility of rowdiness at the Sundowner. If it boiled up odds were that the thunderheads formed around the pool tables, which were in another room, then stormed into the dancehall in the form of a cowboy or two flying through the air. Most often that is where it stopped unless he happened to have friends at the bar. Later into the night the dancers occasionally found it necessary to twirl around a shit-faced loner who'd passed out on the floor and hadn't been picked up yet. The best dancers did this with tremendous style and the barmaids, who were the kind that could deftly weave dollars folded lengthwise through their fingers, barely noticed at all.

At two in the morning on Sunday it was over. The glass pack mufflers roared up in the parking lot and the cat and mouse game of trying to get home with too much alcohol in the blood began.

The Sundowner finally ended up closing down for good a few years back. This hand-scrawled sign was posted on the door:

CLOSED
Too many DUI'S—Everybody in Jail

Beyond the cities and towns and bars and out-of-the-way watering holes stood the forest and range. It was the place we went back to on Monday mornings. Most of the time it was as regular as any other place of employment, but once in a great while there were moments that didn't bear up to the forms of reasoning that we carried in with us.

They were the quirky, statistically irrelevant kind of things that happen when everything in the factory where you work is alive. The moments, and that's all they were, implied the possibility that the textbooks might be incomplete and our thinking flawed if only for the simple fact that we know more than we are willing to admit. It could be that consciousness is weirder than science and that there are alterations that go beyond a beer drunk down at the local pub. It is possible that some *big* medicine really does exist.

I was marking timber on Stoner Mesa east of Deer Creek. It was toward the end of August when the rains are common in that part of southwestern Colorado. A typical day then meant thick, low clouds in the morning with the expectation of rain later on. At the time there was a thin drizzle that seemed to hang in the air, too light to come to the ground.

I spotted a coyote absent-mindedly trotting through the timber about fifty yards from me and stopped to watch him. He looked up occasionally as he went about his business. He kept coming my way and I kept thinking, he has to know I'm here—he's looked right at me. He got closer and closer. I began to get a little nervous. I started thinking maybe I should shout or rustle the brush, formulate some kind of warning. I mean this is a *wild* animal and I am a human. Where is the proper distance?

I kept quiet and the two of us collided. The coyote ran right into my left leg, sort of backed off, then looked up with a kind of "what the hell" expression. He then took off. I started to shiver, which is what I do when I'm cold or scared or both.

There are logical explanations. Maybe he *really* didn't see me. Maybe he *really* didn't smell me. A wildlife biologist I spoke with about it figured that he must have been sick. He said it could have been rabies or any number of other problems. It's all possible, even plausible, and I have to face it—it's the way I *want* to think about it. But I saw the look in his eyes after the inexcusable occurred and I think that coyote was okay. He could have been having a bad day. What's more he might even have been *trying* to make contact. Anyway, statistically it's irrelevant. Just one of the cases that doesn't fit into the model.

I had another run in with a coyote once when I was bow hunting in the La Plata Mountains. This one was running up a trail toward me. When he saw me he freaked out and jumped off a cliff—but I never *heard* him hit bottom. Easy enough to explain. Too far a drop, too much wind, my ears don't work. Never found a body, either. I kinda wanted the skin.

If you think about it anything can be taken as an everyday occurrence. So a bird lands on your shoulder or a coyote runs into your leg. What if nothing more than a funny wind blowing through the timber makes you scared for no explainable reason. . . .

I hunted elk one season when we worked out of the Glade Guard Station. I didn't work it too hard and usually ended up driving out after work and wandering around until I found a nice spot to sit. I tried to pick places where I might see an elk.

One evening I was heading down toward a little draw, actually not much more than a ruffle in the topography that I remembered covering earlier in the summer when I'd been scouting out timber sales. There wasn't much to it, but there was a meadow there that I thought might be a place the elk would come to graze at twilight. It was fairly close to the road and an easy walk.

On the hike the thought came to me that I just absolutely needed to go north, up the draw, from where I wanted to be. It was not the fleeting kind of whim that is easily dismissed, but I still wanted to go where *I* wanted. I fought it, but the idea would not let me be—I headed north and found myself at a little watering hole, the kind that the ranchers dig to catch the runoff for their cattle. It looked like the place to be, even to the point of a big ponderosa pine that I could lean up against to watch the water. That's where I sat.

Just at twilight I heard the commotion of elk working their way up to the water. There were hooves over rock and the thin whistle of a bull. I sat and watched them come into the water and killed a spike bull elk there. The entire time I had sat there it never occurred to me that I wouldn't see elk, that's how strong my feeling for that place was. When I went to dress out the spike I noticed that he had a badly torn up hoof. It was hobbled to the point that I doubt he could have made it through the coming winter.

I have learned to not try to explain the clairvoyance that happened for just a moment that evening. I know that a kind of distance was removed then and replaced for a time with just plain clear vision. I don't let anyone try to tell me that the call didn't go out that evening. I just happened to be a predator in the right place. It was an everyday occurrence.

FIRE

The pictures were all over the national TV news. There was a squad of Forest Service firefighters standing in front of a house somewhere in southern California. The brush for as far as the camera panned was blacked down into piles of smouldering ash. They had saved the house. They grinned into the camera and the ecstatic homeowners hugged and kissed them. Dan Rather smiled.

I noticed that one of the firefighters clutched a small note pad in his hand. It was the standard issue three-by-five-inch tablet. The pages are gridded in blue quarter-inch squares. There are 216 squares on a page and the top is perforated so each page can be neatly torn off if required. The acquisition clerks would know that this is the 1300-18a(3-66). To most of us it is simply known as the Idea book. This comes from the message emblazoned fully across the cover:

*I*nnovate
*D*evelop
*E*ncourage
*A*ctivate
*S*uggest

It is a simple acronym meant to guide us along our way. In the upper right hand corner of the cover are the italized words "Work Safely Don't Get Hurt!".

I know why that firefighter was holding so tightly to his Idea book. From the second he had received the fire call from wherever he was stationed he had been keeping track of his fire time. He had carefully noted his regular hours and the quarter time hazard pay he would get until the fire was declared controlled. There was another column showing his overtime pay that would go time and a half plus hazard pay. There might be a column for night differential.

Time and three-quarters is what can make a grunt's day on the fireline. At any given moment that firefighter who was standing out in the ash and smoke could probably give you an exact figure in dollars and cents of how much money he had coming. It helps *E*ncourage him to *A*ctivate when he is dirty, tired, thirsty, and bored.

I don't know what makes Dan Rather smile, but I do know what makes firefighters smile.

Most seasonal forest fire fighters get started on their home districts. At the beginning of the season there is an orientation that explains how the district is operated. A couple of days is usually devoted to fire training. Firefighters learn how to size up "smokes" and hopefully put them out. The basic notion in most wildfire suppression is that the fire must be deprived of fuel by cutting a fireline around the burning area.

There are hundreds of refinements to fireline theory, but it always works down to clearing away anything that might burn from the fire's path. This means a "line" dug down to mineral soil all the way around the fire. It might be two feet wide or the width of six or seven D-9 cat blades depending on the size of the fire. Other options include taking the heat from the fire by cooling it with water, if water is available. Still, the mainstay is always the fireline. This is sacred.

The tools that a firefighter uses are basic. A shovel to scrape and dig fireline and throw soil on hotspots to cool them and the Pulaski, which is a combination ax and grubbing hoe, to break up the soil and cut small trees and brush. Other more specialized tools like chainsaws, brush hooks, and the rake-like McCleod sometimes come into play, particularly on the larger fire crews.

The working organization of a fire suppression crew is similar whether it is a two- or three-man "initial attack" crew or a standard twenty-man hand crew. The firefighters line out with Pulaskis in the lead and "brush" out the fireline and break up the sod or duff. The shovels follow and scrape the line down to mineral soil. Sometimes a chainsaw man will lead off if there are trees or a lot of brush that needs to come out. Brush hooks may also work along with the Pulaskis. McCleods may be used in conjunction with the shovels.

The idea, particularly in the larger twenty-man crews, is for each firefighter to give the line a "lick or two" then move on and let the next guy do the same. If the crew is "well oiled," they will have built a perfect fireline, down to mineral soil, by the time the twentieth man puts his licks in. If it isn't, either the firefighters up front will do all the work or the ones in the back will. A good crew can build a lot of line quickly because, theoretically, no one firefighter gets too tired.

The entire fire organization is set up to deal with a much broader spectrum of eventualities. The direction is mostly oriented toward the very big, or project, fires. These fires often employ thousands of individual firefighters and an array of equipment.

On a big fire the twenty-man crews might be organized into divisions and sectors. Some of those crews might be hotshot or interregional firefighting crews whose main

function is wildfire suppression — sort of the Green Berets
of firefighters. Specialized engine crews lay hose, supply
water, and protect structures that might be threatened.
Fixed-wing aircraft might be used to drop slurry. Helitack
crews utilize helicopters to transport firefighters, scout out
the fire, and drop buckets of water on hotspots. There
may be overhead teams, Dozer bosses, tree felling bosses,
supply bosses, transport bosses, information officers, recon-
naissance teams, field observers, planners, tacticians. . . .
It has all the beauty of a war except nobody's shooting at
you. Actually, it's amazing that as few people get hurt or
killed as do on the big fires.

Despite the technology of it all the bottom line still
comes down to the basic fact that everyone, either directly
or indirectly, is trying to remove the fuel or the heat or
maybe even the oxygen from the wildfire. And it could be
that the wildfire, on some weird level, is a life itself.
Something to be called by name. Sometimes, "Sir."

Most firefighters don't get their start on the big ones.
It's seldom the kind of blazing inferno that makes its way
to the nightly news. Most often, at least in the Rockies, it
is a lightning strike that has managed to nail a single
pitchy snag. One burning tree can puff up a lot of smoke
and the district will dispatch a couple of firefighters out to
the area. Although the smoke is easy to see from the air
or maybe from a fire tower posted off on some distant
ridge, it can be a different story when you are looking for
it on the ground in the jumble of the forest.

Smoke does funny things when it is gliding across the
land. If it's early in the morning when the wind is coming
down the draws, smoke can carry miles from the source
and layer into a hollow of cold air. When you are new at
it you will think things are quite simple when you see a
concentration of smoke like that and dive off a ridge and
follow it. Maybe after searching for an hour or two you
learn a first lesson — where there's smoke there isn't
always fire.

I remember a burning snag on the Pikes Peak ranger

district in Colorado. The fire tower had spotted it late in the evening and they called us up and told us to be ready to go after it first thing the next morning. There was a new forester on the district and he was going along. He was straight out of college and had all the course work in fire management. His skills on the ground were limited, but the Forest Service figured this was a good way for him to learn—they made him the boss.

We gathered at the office at dawn and headed out. The night before we'd pulled the air photos of the area and found a road that would get us close. It took about an hour to wind our way up out of town and into the area. When we got out of the truck and had a look we saw a small draw about half a mile off that was filled with smoke as far down it as we could see.

This is when we learned that the forester had apparently had some leadership training, too. He took the show over in a characteristic military sort of way. This isn't unusual on a fire because a para-military type chain of command and mode of action is well suited when you have to move lots of firefighters and equipment as quickly and efficiently as possible. On these smaller smoke chasing deals we usually would forego some of the formalities, but we figured this guy was green.

He barked some orders out, we grabbed our tools, and busted double time over to the draw. We didn't talk it over when we got into the smoke and headed down the draw. I knew where the fire was and so did the other grunts with me, but we followed. After about forty minutes we were working hard not to giggle when we left the chain-saw and some hand tools leaning up against a tree. The forester didn't even notice as he raced around searching for the fire.

Finally, we broke out laughing when he asked us where the hell our tools were. We told him they were up the draw toward the fire and his pupils began to dilate. Sweat was pouring out from under his hard hat. We pointed all the way back up the draw. A single ponderosa pine was puffing smoke on a ridge not more than a thousand feet from our truck. He took it well and even choked out a muffled laugh as we headed back. It has happened to all of us and we never told anyone in the office about it.

Fighting district wildfires is the best kind of apprenticeship for a firefighter. Sometimes they turn out to just be a snag, but in other cases it could be three or four acres or maybe upwards of twenty. We learned how to size each smoke up and decide what to do about it. Sometimes if the fire was just crawling along in the duff we took our time digging line around it.

If the fire was hotter and walking or even trotting a little we often decided to hotline it, which meant we went right to where the fire was moving the fastest and tried to get a line around that area. If things worked out it meant that we stopped the advance of the fire at the head and then could take our time finishing the fireline out around the areas that had already burned or weren't burning very hot.

It was good training in the basics of how to safely carry the tools over rugged terrain and pace ourselves for the long haul when we used them. We picked up tricks from other more experienced firefighters, like how to push the shovel with a knee while scraping line and conserve our strength. We learned how to anchor the fireline into a creek or rocks or roads so that we knew the fire wouldn't get behind us.

The details were critical, like making sure to throw any twig or branch that had burned in toward the fire but anything that was totally green outside the fireline. This prevented the possibility that even an ember might get outside the line and start up some trouble. The saying went, "green to green and black to black."

When there was a line around the entire fire, the dirty, boring, and absolutely crucial mop-up work began. On the smaller fires this meant going to every log, every stump, every ember, and every pine needle that had any fire in it and putting it dead out. Most of the time we did it by "dry-mopping," which meant we chopped the embers off logs, then mixed dirt with them and spread them until we could hold a hand to them and feel no heat. We did the same for anything else that held any heat. They called it "cold trailing."

Sometimes we carried five-gallon backpack pumps with water that we used to cool the embers and duff down before we stirred and spread. The backpack pumps, or "piss bags," were miserable and heavy to carry but greatly speeded up the process. We used the same mop-up techniques—it was just quicker with the water. The rule of thumb was that if you didn't know how to dry-mop, then you couldn't wet-mop, either. Just spraying water on a hotspot won't put it out—it needs to be broken up, stirred, fiddled with.

Most importantly we learned that fires and firefighters keep some odd hours. We learned about telephone calls in the middle of the night and sticking with things even though we were going to have to work through the entire night following the light from our headlamps. We learned that you could be more tired than you ever thought was possible and keep digging fireline. We learned how to take care of ourselves and watch out for the occasional fire that the wind would throw into the crowns of the trees where it could take off into an uncontrollable run. We learned that you don't always stop them and that the little fires can get big in a hurry. And that sometimes people get hurt.

All of it prepared us for the big fires, the project fires, where the call might come at two in the morning and before you knew it you were on an airplane heading out to some monster fire in California, Idaho, Montana, Wyoming, Nevada, Virginia...a fire that would have camps with hundreds, even thousands of tired firefighters sprawled out, sleeping on the ground or trying to get to a telephone to let someone know where they were. It can get into your blood.

I would like the project fires for no other reason than the going. Aside from the adventure of it all and the mystery of where you might end up and what you might see and the money you might make I would settle for the simple pleasure of being constantly in motion. During the fire season most firefighters have a packed bag standing by the door or in the trunk of their cars. It's an everyday reminder that at any moment a call might come and you will be on the road.

It is travel stripped bare. There are no photo stops, no educational sidesteps, no chance to get a feel for anything other than the wind over a truck's windshield or an airplane's wing. The idea is to just get there. Your meals are covered. You throw a sleeping bag out wherever you end up when you can't continue. You catnap. You keep moving. I was born with a feel for it.

The fire crews that mobilized for the great California fires in the fall of 1987 did it so well that the Army came by and asked how it was that they could get five or ten thousand people and their equipment on the road and headed out in just a day or two. They'd been trying to find the answer for years. The Israeli and the Swiss armies and apparently the American firefighters know the answer. It's simple enough—they have all their gear at home, ready to go, actually just waiting, almost hoping to go. The Army tends to draw the line at M-16s stashed in every soldier's home.

I was called to the California fires that year and in a lot of ways they were like most of the other big project fires I've worked on. There was one exception, though. We drove a strike team of five fire engines from Colorado to California. It seemed like most of the engines in California were the big jobs and the need for the small, mobile, four-wheel-drive type of initial attack "pumpers" that we use in the Rockies was recognized. We left Durango, Colorado, at two in the morning, two firefighters to a truck, one driving and one wedged in that dreamy zone this side of unconsciousness—the place where the little green men live—and we drove for thirty hours with a three-hour nap in Ely, Nevada.

I know fatigue much better than I know the other

phantoms in my life and I can thank firefighting for the friendship. The rulebook says they can work you for the first 24 hours straight when you get the fire call and after that you're entitled to 12 hours on duty and 12 hours off. It seldom works that way, at least in the early stages of a big fire. Figure some stretches at 18 or 20 hours a day, figure 100- maybe 120-hour work weeks if things get ugly. Figure on no relief. Figure on making exhaustion a friend rather than the enemy. Make it a give and take relationship and hope that you're "on" if things get hot. The little green guys dancing across the road, or the fireline, or in the mess line are all right, but draw the line at the little blue guys.

I was working night shift on a big fire near Fairplay, Colorado. My job was simple—just make sure that a Pacific Marine pump located by a small stream kept the water coming to the firefighters up on the line. There wasn't much to do other than gas it up every few hours and listen for any break in the mechanical hum that might mean things were going wrong. After three or four days of it I got to where I could doze next to the pump and wake up instantly when it started to sputter for lack of gas. Sleeping off shift, during the day, wasn't going well. The fire camp was noisy and too bright. We were getting three or four hours a day.

One night I figured to break up the boredom and walked over to talk to another firefighter that I'd noticed rummaging around in the back of a nearby pickup. It turned out that all that was there was a broom, stuck up in the bed of the truck by its handle. I don't think I talked to it for very long before I realized. . . .

The language conveys something of a chromatic scale—tired, dog-tired, fatigued, bushed, beat, worn out, weary, exhausted, goneness, dead tired, spent—but these are only the notes in a fugue. You can come to understand the orchestra, the composer, and even the performance.

They put us up in a motel when we pulled into Redding, California. It wasn't part of the plan. The dispatcher had figured to head us straight up to the fireline, at least, until our strike team leader told him that the crew wasn't going to move a goddamn inch until they got some sleep. We'd been seeing the little blue guys.

At four the next morning we left for the fire. We made the district ranger station in Hayfork, California, at dawn, just in time for chow at a fire camp that had been set up nearby. The place was chaotic. A huge storm full of lightning, but no rain, had torched eighty percent of the district. They didn't even know how many fires they actually had burning. They dispatched us to another fire camp farther to the west. Smoke filled all the valleys.

The fire camp had been up in the mountains, but when we arrived they were in the process of moving it to a huge field near Hyampom, California, which they figured would be big enough to hold the expected two or three thousand firefighters who were on the way. We grabbed the best campsite we could find in the new fire camp and set up shop. Word was that we would be pulling night shift. We left for work at six that evening.

At first it was easy duty. We'd been charged with protecting a fireline that had been backfired on the previous night shift. Backfiring is a form of wildfire control that you see a lot of on the bigger fires. Generally a fireline is dug, or in our case a road is located that is in a position ahead of the direction that the wildfire is moving. Controllable fires are deliberately set at the fireline with the intention of burning out the fuel ahead of the fire. When the wildfire meets with the backfired area it is deprived of fuel and theoretically stopped. It's a way of creating the *very* wide firelines necessary to hold huge wildfires that can jump over any line dug by firefighters or bulldozers.

A lot can go wrong with a backfire. If the wind changes and it gets out of hand it can jump your fireline and you're right back where you started. If things go right the backfire can eliminate the fuel from the oncoming fire and even change the direction or force of the fire's convection column and turn it back into itself. Most backfiring is done on the night shifts when the humidity is a little higher and conditions don't favor the kind of blowups that can occur during the day shifts.

They like having pumper trucks like the ones we'd brought from Colorado to follow the firefighters who are lighting the backfires. If the thing gets out of hand and spots over the fireline, the water is available immediately to put it out. Spot fires are the firefighter's enemy. If they get out of control not only do they mean that all the work put into the fireline is worthless but that you may end up having wildfire on both sides of you. This isn't considered sporting.

The section of line we were patrolling was in good shape. The backfire had gone as planned and there were just a few hotspots, here and there, that we figured were too close to the line and put out. It looked like an easy shift that held the potential for a catnap. An experienced firefighter will always sleep when he can, because he knows that at any minute he may be called on to go full out at one hundred percent if something blows up. We laid back.

It didn't last. Around one in the morning they dispatched us to a couple of spot fires that had jumped the line on another part of the fire. From the beginning it was what firefighters call a watch-out situation. In fire training they give you a little book with a skull and crossbones on it. The title is *Fire Situations that Shout Watch Out*. There are thirteen of them. What they are trying to get across is that if you don't watch out you may get your ass burned.

As axioms for survival go they are pretty simple. Take number seven—"You are in country you haven't seen in the daylight"; number ten—"You are getting frequent spot fires over your line"; number eight—"You are in an area where you are unfamiliar with local factors influencing fire behavior"; number one—"You are building a fireline downhill toward a fire." I like number thirteen—"You feel like taking a little nap near the fireline." All of these and a few more applied. There was that wonderful jolt of adrenaline that watch out situations can provide. The kind that can get you through an entire night shift. We checked for our fire shelters.

Yes, the Fire Shelters. They are the neatly packaged outfits that look like an aluminum pup tent when deployed. The idea is that a firefighter who has run out of options can find a spot relatively clear of fuel, hopefully a purposely created wide spot in the fireline called a safe zone, set up his fire shelter, and get into it with his feet pointed toward the oncoming wildfire. The aluminum and fiberglass lined fire shelter will reflect the intense heat of the fire away from him while he breathes the cooler air an inch or two above the ground. He can crawl around in the shelter to get away from any unbearable heat if necessary. He must hold the shelter down tightly when the fire and its accompanying cyclone of winds goes over him. The shelters are proven and there are a number of firefighters alive today thanks to them, but it still doesn't keep us from calling them "Shake and Bakes."

Things *do* happen. Years before I'd been on a project fire in southern California. The fuels there are flashy to the point of being explosive. I was on a hand crew that was backfiring a section of line out. The squad doing the firing was using flare guns. The key when backfiring with flare guns is to shoot the flare in from the fireline twenty or maybe thirty yards. When it ignites the fuel the fire doesn't have enough time to get a head on it before it meets up with the preconstructed fireline and can be controlled.

This particular squad got to screwing around and was shooting the flares too far down a steep slope of heavy fuel. The fire took off and overran them. I managed to run down the fireline to safety, but some of the crew had to deploy their shelters. It was over as quick as it started and we ran up to them. Everyone was okay except for one firefighter who'd had his sleeves rolled up. The shelter, which he had to hold down with his forearms, had gotten so hot that it burned his arms. It wasn't as bad as it could have been.

Another time in southern California I was on a twenty-man crew along with four other twenty-man crews on a narrow ridge away from the main fire—a watch out situation. We were there awaiting instruction from Command. The main fire shifted and came our way. The crew leaders radioed for instructions and were told to stay put. The main fire, which was still a couple of miles off, kept coming, but we could hear it even that far away. People started getting edgy. The crew bosses called in again and were given the same instructions. The fire began to run.

That's when the crew bosses began looking for spots to deploy our shelters and came to the conclusion that the country was so steep there wasn't enough room for the 100 firefighters on that ridge to put out their shelters if they needed to. We were backed up to a steep, almost vertical slope on one side. The fire was boiling when they called Command and requested a slurry drop to try and slow the fire so they could think.

The planes came in low and made their drops. There was no talk because even us grunts knew things were getting tight. Less experienced crews might have bolted and that would have been the kiss of death. The slurry slowed the fire just a little, but it was enough for us to pull up a hose lay that was on the fireline, tie it into whatever we could find, and go hand over hand down the steep slope to safety. That's when we started the banter, yelling at the crew bosses to call in for 100 pairs of clean shorts.

Those are the exceptions. Most of the time the firefighting is boringly routine. If you are going to get hurt it's more likely to occur when you're flying, or being transported by the National Guard, or by the sudden randomness of a huge burned out snag that falls out of nowhere. The snags may be the most feared because there is absolutely nothing you can do if your time is up. It takes some getting used to.

Those spot fires when we were on the pumpers in northern California? Well, we didn't do too well. They got away from us. Another crew came up but refused to help because they thought the situation was too dangerous. We pulled back when the flames started racing into the crowns and lit backfires to try and slow it, but it was too close to morning and the sun heated things up. The whole thing blew up on us. We went back to camp after being up twenty-five or thirty hours. As they say, "That's show biz!" The pay really is the same.

Eventually, the crews did get a handle on the fires, which were collectively called the Gulch Complex. Our lives turned into an endless string of patrols and backfires—all night shifts. The smoke hung in thick and once or twice the carbon monoxide took its toll in the form of nausea and delirium, but we just figured it was fatigue. One shift it was so bad that when a division boss came to talk to one of our engine bosses he found him completely incoherent. The division boss asked, "How long have you people been out here?" We couldn't tell him. We became the walking dead.

I remember the rest of that time in California more like the two-minute sound bites you get on the nightly news. It is all in fragments and the story line is hard to find. I remember sitting around a warming fire that Frankie Maestas and I built one cold night on the fireline and talking about how we would stop at Reno if they ever let us go home. It struck us as funny and we laughed until the tears rolled down our cheeks and we slid into that deep, dark kind of fatigue that even scares the little blue guys away.

I remember getting released from one fire only to be dispatched to another. Fire camps that looked all the same in the Seiad Valley, Happy Camp, North Elk, Norcross, Forks of the Salmon...and endless night shifts. There was R&R for a day or two in Yreka, California, and the feeling that I would never get enough sleep. Our strike team leader was going nuts trying to get us relieved. One morning I looked over at one of the firefighters on the crew and I realized how bad it was. There was no light to be found. Her eyes were dull. I figured that we all must have the thousand-mile stare. We were sick and coughing from the smoke.

Finally after thirty-five days they sent us home. We

made it to Reno, but when the leggy bargirl came over and asked what she could do for us, we just stared into space.

"Somebody say something," she said. We told her we'd been on the fires for a very long time.

The crew split up after Reno and we all drove home our separate ways. We'd had our fights and reconciliations the way all crews do on a long run, but most of it was forgotten. We were family whether we liked it or not.

I ended up driving alone for the final leg of the trip from Grand Junction to Durango. Try it after you've been with ten other firefighters day and night, good and bad, for thirty-five days. It's a new kind of aloneness. The best month in Colorado is September and it was over, but the air was clear and some of the aspen were still brilliant and yellow as I drove over Red Mountain Pass. It was nice to see mountains that I knew. It was a lot like home.

I read later that the Fire Siege of 1987 involved 22,000 firefighters in California and Oregon. In California 775,000 acres burned. Enough timber was burned to build homes for a city the size of San Francisco. Fire suppression costs in California alone were over $100 million. An interesting footnote gave the results of a study on the crews that pulled night shifts on the fires—they averaged three to four hours of sleep a day. Ten firefighters lost their lives.

And none of it, with the exception of the lost lives, would hold a candle to what was to happen in the Yellowstone National Park area in 1988.

Wildfire fighting, at least the grunt work that most of the seasonals do, is a job for young men and women. Needless to say, the hours are long, but the money *is* good although there comes a time on the big fires where even the money doesn't matter.

After twelve years of firefighting I've gotten to the point where on any project fire I go to I'll see a firefighter that I've met on some other obscure jumble of topography that

149

was burning. In many cases it could be that the wildfire was the best thing that ever happened to the landscape. But we were there anyway, whatever the politics of nature are.

I could say that after having worked on somewhere close to a couple hundred wildfires, both large and small, that I'd had enough and it would be almost true. But there is another part of me, the part that just lives to be on the move, that is always looking for the next fire call, the next column of smoke over the next ridge, and wondering what that fire will be doing. Will it be crawling in the duff or roaring up some ridge? Will it be the kind of fire that charges me full of adrenaline and wonder? Will it be the kind of fire that has a life of its own?

I'm still no different than all the other grunts. I'm still in it for the rush and the cash.

MONICA

We were living in northern Idaho the first time she threw my ass out. It was late enough in the winter that I had already begun to think about spring, but I was a little premature. Getting kicked out of my own house brought back the meaning of wind chill. I heaped my stuff into the pickup and we only had one small disagreement over who would get the tent. I took it because I was the one heading down the road.

We'd met in a botany class seven years before. Monica was the prettiest thing I'd ever seen. On our first "date" I talked her into going up into the high country to collect plants. I was after an edible called bistort and knew a field where it grew off the Peak to Peak Highway west of Boulder. We ended up becoming lab partners and the rest was simply romance. We got married a year later.

I didn't blame her for throwing me out. I had taken to an itinerant kind of life as soon as I got out of college.

The deal was I wanted to work outside. I was gone from home for weeks at a time and never made much money, but I always came home with tales of rivers and mountains without end, wild places and dreams. I think more than anything she liked hearing the stories. I could see those places dance across her eyes when we sat down and talked after a week apart. In the off season and on weekends we took off to the mountains and deserts—sometimes into Mexico. Once we moved to Eagle's Nest, New Mexico, on a whim I had to get away from it all. Besides, I liked the name.

I believe that she thought I'd grow out of it, but each year I kept signing on for another season and ending up off on some ranger district in the middle of nowhere. It didn't seem right to her that a boy from the suburbs with a university degree would choose a drifter's life roaming the woods at low wages forever. Finally, I think she might have gotten tired of the stories and there we were saying goodbye to each other.

I headed south through Kendrick to the Clearwater River. There is a good gravel road there that drops 2000 feet down from the prairie to the Cherry Lane Bridge over the river. It's a long one-lane bridge with a stop light on either end. If someone else is on the bridge the light turns red and you have to wait for them to clatter across.

The bridge is also a good place to look down into the river for the steelhead. There are some good holding spots where they rest up before busting upstream through the faster water. When I fished the river I used to run out onto the bridge, spot the steelhead, then run back down to the river and try to cast a fly to them. Once I had a big, strong fish take the cast. I was using an old reel that was bent out of true and the first run the steelhead made literally blew it apart. The fish broke off and it was the only steelhead strike I ever had in Idaho. That day I was heading out of town I wondered if there were any early spring steelhead down there finning the water and saving their energy, but I didn't stop to look.

If the urge to fish is inherited, then I have inherited it; if it is a disease, then I have caught it. All I know is that fish swim through my dreams. I think about the hordes of steelhead that once swam from the Pacific up the Columbia River to the Snake River then finally into the Clearwater. There was a tribe of them that made their way to the North Fork and the fishing there was storied.

The centuries-old cycle of that migration ended when the Dworshak Dam was constructed. That great run of one of the largest strains of steelhead in the continental United States found themselves cut off. Each year the collective memory of an entire population of fish has receded. They are being replaced slowly by hatchery stock brooded at the foot of the dam, but there are still some wild fish that nose up the mile or two of the North Fork that remains. I headed up the river toward Lolo Pass.

I have given up on trying to understand what fish think. It could be that those North Fork steelhead thought nothing about the end of their imperative, but I believe that on some level they understood that somehow, some way, something was not right when they came up against the concrete of the dam. I was in for a sad passage over Lolo Pass myself.

Lolo Pass has never been easy. When Lewis and Clark were coming out of what would become Montana into the future Idaho, things got rough for them at Lolo and farther west until they made their way down to the prairies. Even today the road over the pass to Missoula is narrow and windy in spots. It's been paved in the past twenty years or so, which makes a difference, but still, from Lowell, Idaho, east for eighty miles there are no towns or services. In the early spring, on the lower stretches of the road, there is a good chance for black ice that looks like a wet spot on the road but is really a deadly sheet of ice. Higher up the snow can be very deep.

I drove past Lowell and into the forest. At first it was open but got thick to the point of opacity as I gained elevation. It seemed like the very denseness of the forest closed all things out. A fog came into the trees along the way and the radio began to crackle and sputter as the stations faded.

A tree's life is not as simple as it might appear. At first, when I started working in the forest I thought I'd just about figured them out. After I'd been around awhile I realized that there are shadows of great mystery that surround the lives of trees. It could be that all those trees on Lolo Pass represented what is still in life. Maybe their vision was one of the stillness *between* the cars pushing their way up the pass. The knowledge in that forest might be of what occurs between the motions. It could be like what happened between Monica and me. Something between the motions.

It was snowing when I crested Lolo Pass and came into Montana. The forest there begins to open almost immediately in response to the lessened rainfall, which is wrung out of the clouds when they wash up against the Bitterroots on the Idaho side. There is a kind of translucence in Montana that isn't apparent in northern Idaho.

I had called ahead to my friend Paul Krupin, who lived in Salmon, Idaho, and explained my situation. It wasn't that it was unusual; men and women split up all the time. It's just that universals of this nature seem so much more *universal* when they happen to me. Paul volunteered his couch and a roof over my head until I got back on the feed.

Paul and I had met when we lived in the same dormitory at college. We had a strange chemistry between us, but when the equation was reduced to its simplest terms we ended up friends whether we liked it or not. He must have been the smarter of the two of us because he ended up with a permanent job as a hydrologist with the Bureau of Land Management in Salmon. He had been advancing through the ranks while I was still wearing blue jeans and sleeping in the back of pickup trucks.

There is no way to drive to Salmon, Idaho, that isn't pretty. I'd come down off Lolo Pass and headed south through the Bitterroot Valley, which is one of the prettiest.

It's the country Norman Maclean couldn't get out of his system and wrote about fifty years after he'd worked as a Forest Service seasonal there. For a trout fisherman the Bitterroot is a thick thread of mystery that verges on the romantic. I drove by it and wondered if I'd ever be in the mood again.

After Lost Trail Pass the road drops back into Idaho and the Salmon River. The Bitterroots jut up to the east and Salmon isn't far down the road. I got to Paul's in the afternoon. For the rest of the day we circled around the issues of our lives like a couple of vultures. Paul was lonely in Salmon and wanted a girlfriend and here I was with the broken heart. We never quite zeroed in on the meat but just hovered and circled. Then we got drunk. The next day he took me over to some hot springs in Montana and let me soak. I was in bad shape.

My condition slowly improved. I started going down to the river on the warm mornings after Paul went to work. I just walked it and smoked cigarettes because I was too strung out to fish. I knew that there was the chance, slim as it might be, that some steelhead might be coming on. I thought if I could just see one, maybe finning in a deep clear pool, resting, that I too might come to rest. A big, true sea-run steelhead held value for me then. I thought if one of them could make it all the way to the Bitterroots, damn near into Montana, that there could be hope. I load the fish in my life with a lot of baggage.

Paul was doing as good a job with me as my own mother. In fact, he'd talked to her on the phone several times. I believe they had decided that the best course of action was to keep me busy. Paul's plan was to put me on to a new girl, but it was too early. It might be that other men recover quickly, but I wasn't one of them. I still looked at the whole affair as a burst of turbulence, something I'd bump through, then come in for a safe landing. I believed that Monica and I still had unfinished business and that we would come back together. More than anything I missed the way she talked about books, not just as

ideas, but as lives affected by the size of print, kind of
cover, binding, and harshness of use.

Paul took me to a dance Friday night. It was a
gathering of the people from his office and some other
government workers in the area. I wasn't happy to be
there but thought I could get through it if there was
enough beer. A woman was there who was in the process
of falling out with her husband and it might have been we
could have helped each other. The rumor was that the
husband was losing interest. I think she was looking for
the jolt to power her on her way, whichever way that was.
She pulled me onto the dance floor and tried talking—Paul
had let her know about my predicament. She wanted to
know how to tell when things have turned for the worse.
In my case it had been as simple as a finger pointing
toward the door. I couldn't help her.

I started casting a fly line on the Salmon River soon
after that. I worked places where I knew there would be no
fish because it was my goal to just tighten up my loop. I
had never cast well. One morning when I was driving out
to the river I saw the girl from the dance. She was dressed
in a nice wool skirt and blazer and I believe she had
nylons on. She had a suitcase and was heading toward the
bus station. I waved.

I began to get on Paul's nerves. I could see that he
was in danger of crashing, too. He was smoking more than
he liked and also drinking some. We decided that I needed
to move on. Get out and experience life. I called John
Gierach in Colorado.

John and I went back a long way and he was the most
competent man I knew when it came to fishing and women.
I had hoped that it wouldn't come down to this, but I
would have to take the cure at Dr. John's. He said I could
live in his attic for as long as that took.

John's place looked the same as it always did. A
scattering of cats watched me wide-eyed and fled for cover.
There was a pile of wood heaped up in the driveway. A
flock of Plymouth Barred hens scuffed around the gravel.

MONICA

The chickens had formed part of a great experiment when John had decided to try and raise dry fly hackle. The results had not been encouraging, but there had been profit in eggs and the soft Grizzly hackle. The problem had been the roosters, which are the source for the stiff, quality hackle needed to float a dry fly. They were a miserable bunch of customers taken to pecking each others' eyes out and other associated acts of terrorism. John, in response, developed the habit of walking out back and shooting them when things got out of control. The hens were all that was left and John had civilized them to the point of being pets. He called them "the girls."

I was met at the door by Strider, the Dalmatian, who could be vicious to visitors whom he didn't know or remember. He couldn't place me, although we'd met, so I called through the screen door to John. He rumbled down the steps from the attic with a bag of dogshit he'd just cleaned up in anticipation of my arrival.

The inside of the house was warmly chaotic in a utilitarian kind of way. One entire room was set up for fly tying with animal skins, furs, and feathers scattered out among vises and hooks. The living room was a kind of demilitarized zone wrapped around a Franklin stove. There were a couple of chairs and a couch that had been shredded by the cats and dogs. The idea was that all the creatures that lived in the house would be equal, at least in the living room.

Another room, which John called the lab, was where he made his living as a fly-fishing writer. It was covered in Kodachrome slides, manuscripts, typewriters, and in the center there was an ancient Chandler-Price letter printing press where he had once printed poetry books. This was not a demilitarized zone and the kitties were actively encouraged to be on good behavior when they crossed its boundary.

There were artifacts of the women who had come through John's place. An afghan here, a plant in the win-

157

dow, a dish or two in the kitchen, and a vacuum cleaner gathering dust in the corner. Strider and a few of the cats also came under this heading. John had a way with women that I'd held in reverence since we first met. It could be as simple as sex appeal if there is such a thing because he is a handsome man, but I took it as a kind of mysterious force. He simply liked women and they came into his life with the regularity of migrating birds. They left that way, too. I coveted his gift and in our younger years simply watched the miracles that came through his door, too awed to speak. I had always been the voice in the corner that called out for journeys into the high country and off to the desert. John often called back saying he'd be along in just a moment when he was finished with what he was doing.

He never mentioned my predicament, never asked for the sequence of events, details, or extent of injuries. His attitude was that I had been in a wreck, more severe than any crash of automobiles or slips or falls, and that immediate treatment was required. The way he saw it I had been hurt and it was necessary to get me over some fish as soon as possible. He had me on the water the next day.

John started me easily on the bluegills at Sawhill Ponds. As fish go the bluegill is one of the more forgiving when it comes to skewed casts or flies that don't match anything that the fish might eat. They come anyway like a trusted dog. This doesn't mean that they don't deserve respect — once they are hooked they can plane against the line with their saucer-shaped form and make out to be a bass twice their size.

Aside from the bluegills, Sawhill Ponds had a calming effect. They are a series of mined-out gravel pits that filled with water, and over the course of time the cattails and tules came in and with that the birds set up shop. Their kind of scrabbling, no holds barred world was thick enough to get lost in. The incessant chattering of a marsh is the white noise of the natural world.

In the realm of high-powered fly-fishing for trout, salmon, bonefish, and tarpon, John had pledged his alle-

giance to the bluegills of the world from the beginning. I
think he saw them as the working class heroes of the
freshwater. An everyman's fish. And it was the ponds
themselves, the entirety of it, down to the smell of the goo
that we sludged through as we waded from pond to pond
that drew him. There is an opulence to bluegill country
that reminds me of the days when cars had the big tailfins
and gas was cheap. I believe, even now, that John could
have driven those big gorgeous sweethearts with class,
which is not an easy thing to do.

We had another, more basic reason for being at Sawhill. Neither one of us had much money and a stringer of
bluegills, especially the bigger spawners, was an enrichment
to our burrito laden lives. Each evening we came in and
scaled and cleaned and filleted our catch, then dredged the
meat in cornmeal and fried it. Freshly caught bluegill is
ecstasy to eat and it did not hurt to have some of what
there was at Sawhill sticking on our bones.

John brought me to trout through the back door when
we started fishing the evenings on McCall Lake. Each
spring the wildlife division stocked the lake with rainbow
trout until the water temperatures rose too high to support
them. After that the fishing went to bass, crappie, and
bluegills. The hatchery-bred trout rose crazily to the flut-
terings of the insects that skimmed the water when the sun
was low. They were not difficult to catch on soft-hackled
flies cast near them.

There was also always the chance that we might tie
into one of the better bass or crappies that had wintered
over, or best of all one of the ancient, used up brood trout
that the division threw in to sweeten the pot. These were
the large four- or five-pound trout that had been used for
eggs at the hatchery but were past their prime. For the true
trout fishermen these stockers were like a walk in a city
park. Real trout, like an extended trip into remote wilderness, were the ones that took care of their sex lives on
their own and reproduced naturally in places like the South
Platte River, the Frying Pan, and the thousands of other

little twists and turns of the mountain stream.

I did tie into a brood trout one evening at McCall. When I got the strike I knew that it was more than an eight- or nine-inch stocker, but thought it might be a bass. I finally got the fish in close and saw that it was a great trout. I yelled over to John and somehow pointed my rod at the struggling fish and it shook the fly out. I remember that fish because landing it was going to be my cure, a quantum leap of cure for a love-wrecked angler.

There can be a single fish, now and then, that grows out of proportion to what it was truly meant to be, almost like a flaw in national policy that leads to war. I still remember that trout now, ten years later, and wonder what difference it would have made.

John got me a sporadic, minimum-wage job helping him package fly tying materials for a wholesaler in Boulder. We spent our hours upstairs from a butcher shop putting feathers and fur into little plastic packages for sale to retail fly-fishing stores. After work I started fishing tiny flies to rising brown trout on the St. Vrain River across from John's house. These were tough, wild little trout and I began to get my chops back.

We decided it was time to head down to the South Platte River. If a fisherman can ever have a river, not so much in the sense of owning, but more in belonging there, then the South Platte was mine. I'd been having an affair with that water, on and off for years. I remember being so crazy about it that on the day we came as close as we ever would come in my lifetime to a total eclipse of the sun, I hustled up there to see what the trout would do at twilight in the middle of the day. Tenderness was never gathered easily on the South Platte, especially in terms of the wary trout rising to my dry flies, but what I have received I have kept.

John and I caught some trout that first trip, but it was the second or third time where I just knocked them out. Whatever had been driven from me was restored. I had also been calling Monica on the telephone and we had decided to get back together. She had said she was having a rough time without my stories and I missed the steady stream of books that she brought home to me like so many trout rising to the evening hatch of mayflies.

I signed up for a season on the Clearwater National Forest at the Musselshell Work Center and threw my things back into the pickup. John beamed as I got into the truck to head back to Idaho—not so much for the choices that I had made, but for the cure that had been effected. You will find this in all great fishermen who can handle a car with tailfins.

Monica and I left Idaho as soon as the Forest Service laid me off in November and we moved to Colorado Springs. Kent Brekke took me under his wing there and put me to work in his fly-fishing store. I started writing some of my stories down in my spare time and took up with the South Platte River again. Monica freelanced as a librarian. A job came up out in the timber on the San Juan National Forest in southwest Colorado and I waffled around just a little until Monica happened to land a library job in Durango and we were off.

We were happy in Durango. The town was small but still the hub of commerce in its sphere. It was also a kind of refugee camp for those people who had come to the Front Range of Colorado looking for the romance of the West and burned out on the urban sprawl of Denver, Colorado Springs, and Fort Collins. There were former engineers, computer scientists, and accountants working happily for minimum wage as waiters, busboys, and hamburger cookers. They had escaped and brought with them a pleasant confusion of mountain bikes, jogging shoes, backpacks, and skis. And there were places around Durango to use all of them. The Weminuche Wilderness was on the doorstep to the east, Purgatory ski area was up the road, the Canyon country of Utah just three hours away, and the dignified glitz of Santa Fe four hours to the southeast.

I went to work for the Dolores ranger district of the

San Juan National Forest. It was about forty-five minutes west of Durango. Monica was the map librarian up at the college. My job meant a lot of spiking out that Monica had come to accept, if not like. Her dad, Sam, had been a soldier his entire life and away sometimes for years at a time, so it wasn't that she didn't understand a man being away from home. I swung it so that I worked ten-hour days, which meant I was gone only three nights a week. In my off time I managed to get friendly with the Animas River, which ran through town, and the wonderful San Juan River in northern New Mexico and its football-sized rainbow trout. I killed an elk out of camp near the Glade Station at the end of my first season and we were set for meat.

The next season started slowly when a number of spring snowstorms hammered the San Juans. We were under the gun that season because the district was contracting out thousands of acres of tree planting on the Glade. When the weather finally broke I headed out to what I knew would be four or five straight seven-day weeks of unrelent-ing twelve- to fourteen-hour days of tree planting inspec-tions. I understood what I was getting into, but the overtime pay would be enough to pad a winter of unemployment.

About the third week out spring overtook the Glade. The grass was greening up and we were able to work in tee-shirts. I'd been following a rag-tag, rough-shod gang of about twenty tree planters for most of that time, but they were doing good work so my job of inspecting was easy. The long hours were beginning to show on me, too. I was a little rough shod myself, having not shaved or cleaned up for a couple of weeks.

One morning I noticed Craig Yancey, the boss of the timber shop on the district, and my friend Dan Rael walk-ing quickly toward me. The supervisors had a habit of coming out to check up on things when the weather was nice so I figured they were out to see the sights. When they finally got up close enough Craig didn't mince words. He told me that my wife had been in a serious car wreck and was in intensive care at the hospital. They were there to drive me back to my truck at the guard station. We didn't say much on the way.

At the station I told Dan that I'd just run in and get my things and be off. He grabbed my arm and looked at me. His eyes were like some deep pool in the river where the bottom is visible but you know that it is somehow distorted, maybe too deep or more shallow than you perceive, that something is off. He said maybe I had better just get in my truck and get going. That's when I realized that things had gotten tough for me.

It was a three-hour drive to Durango. On the way out I passed near the place where I'd killed the elk the year before. I didn't pray to God, or Buddha, or any of the big league, but for some reason I prayed out loud in a rattling chant to the dead elk I'd dragged up onto that road the year before. Coming off Hesperus Hill into Durango I heard over the radio that Monica was in grave condition.

When I got to the intensive care unit Sam was already there. The accident had happened the night before. He was in bad shape when he told me that the doctors said that his daughter might not make it. And I flew into a rage. I grabbed him and yelled that he would never talk that way around me again and started kicking shit around the intensive care ward. I believe the nurses looked over and wondered what had the bearded, dirty mountain man so pissed off and wondered if sedatives might be necessary. I was pissed because I'd been out in the boonies where no one could find me and Sam had made it from Denver to where I should have been before I could get there. It's an old story that is, sadly, the price you pay if you work in faraway places. I was pissed because I was tired and I was scared and it was business as usual in the intensive care room.

Sam and I camped out in the lobby for the next fourteen days while Monica was in intensive care. We slept on the couches and ate in the hospital cafeteria. Every hour they let us come see her for five minutes if she was awake. She made it over the hump after the first three days and regained consciousness the next. Sam was sixty-eight then and a combat veteran of three wars. One late night he talked in a voice that was as tired as any I had ever heard in my life. He had remembered being lost deep in the Burmese jungle during a battle in World War II. He called it the FEBA. The Farthest Edge of the Battle Area.

After things brightened I talked with one of the guys who had been on the ambulance crew. It had been a single-car wreck. Monica had been thrown through the sunroof as the car rolled. Her elbow was broken in five places, the ribs on one side were broken, the lung punctured, and on arrival at the emergency room her heart stopped. She had asked if anyone else was hurt before she became unconscious. The ambulance attendant said that they hadn't been doing very well that week but on Monica's accident everything had come together and they'd been hot. It all fell into place late that night like the glide of a jazz saxophonist into the best riff of his life.

We put our lives back together. Sam went back to Denver. Monica recovered fully, to the surprise of a number of doctors who hadn't expected that much. I asked her what it had been like in that place she'd been. Were there tunnels leading to light, angels, meetings with old friends? She said that she hadn't seen a thing, it was simply darkness, which of course would have made the shadow of the elk I'd chanted to hard to detect. As for me, I never forgot the bunch of flowers that was sent to the hospital by a gang of rag-tag, rough-shod planters out in the boonies.

I started getting more and more work from the Forest Service and my seasons grew longer and longer until finally I worked a two-and-a-half-year stretch straight. In the summers I headed to the field, while in the winter I did map work and gazed out the windows. We managed to get enough money together for the down payment on a house and moved into our own place. The extra work cut down on my fishing.

About seven years after northern Idaho she threw my ass out again. We'd grown apart in a way that was as simple as two trees that were too close to begin with but with age and height had spread their crowns in different directions just to get more light. I had more stuff this time and it took two pickup loads to move it to a small A-frame cabin at the Twin Buttes Trout Ranch off of Lightner Creek west of Durango. I'd been told that it's never as bad after the first time and that seemed to be true, but I wasn't singing in the rain, either. When I'd stayed with John the time before, I'd asked him how he did it. I said he'd had plenty of girlfriends that had come and gone and how was it that he got through. He asked me to please not talk the way I was talking because he didn't know any more than I did. "I just know what's coming now," he'd said. I'd seen it coming, too, this time.

I spent my time tying flies in the evening and started fishing more. I hit the Animas River and the San Juan and the Dolores River. I hiked into the smaller, quieter streams in the mountains. Sometimes when the owners of the Trout Farm went out of town they left me in charge. It was then that I had to get up early and clean the drains on the big runs of growing trout that were used to stock the small ponds on the farm where the tourists came to fish in the summers. Each morning when the trout saw me coming they swam over in swarms expecting to be fed. They weren't wild, but they were family.

Toward the end of September I made a weekend trip to the Dolores River. A big dam had been built there when I worked on the district and the cold tailwater below it had developed into a first rate trout fishery in a canyon that had been too warm for trout to survive in just a few years earlier. The fishing was delicate to trout that rose to dry flies in the clear, low, autumn water.

I ran into Neill Peterson over there and we camped together. He'd fallen in love with the Dolores from the moment the planted trout had begun to grow and rise to the surface. He'd take off from Durango and spend weeks camped by the river while he spent full, long days working over the trout. I'd fished it some, but the Dolores was flowing in Neill's veins. He knew I was having a hard time and decided to let some of his secrets go. We walked the river from top to bottom and he showed me every one of the runs and holes that had up until then been his alone. And we caught trout and sat by the fire at night.

Neill left for home and the last evening I was on the river I fished to rising cutthroats in a deep glide that ran against a big stone cliff. This was where Neill's biggest trout lived and they were rising, sipping Pale Evening Duns off the surface. I took a few and toward dark tied into a big, hard-boring cutthroat. I fought the trout for twenty minutes before I could measure it at twenty inches and let it go. I stood up and yelled.

"I'm back, goddamnit, I'm back!" And it echoed across the empty canyons.

Late that night when I got back to the cabin on Lightner Creek I was called out to fight wildfire in California. It turned out to be the biggest fire bust in that country for the past fifty years. I spent thirty-five days chasing smoke and somewhere in the middle of an endless night shift I began thinking about Monica. I missed her and figured that we still had unfinished business. I started trying to call her at the few times I could get to a telephone, but I never connected. When I came home I called again and she said, "Where the hell have you been? Why didn't you call?" The rest was simply romance. I was back.

It might not seem right to throw a love story in with so many fish. We have a stance in our culture that says anybody past the age of Huckleberry Finn who fishes must not be taking things seriously. The implication is that permanent irresponsibility might follow, which could even lead to vagrancy or at best a seasonal kind of life. This may all be true.

It could also be true that love is very much like a trout on the end of a line, the trick being to never land it. Just play it. Just play it. . . .

FULL CIRCLE

I'm sitting at the orientation for seasonal employees on the Boulder ranger district. It's June 6, 1988. It's been fifteen years since I've lived in Boulder, Colorado. In a lot of ways this is where it all started, at least for me.

Boulder's not exactly the same. I haven't seen many longhairs driving around in beat-up old pickup trucks or VW microbuses—but there are still a few. There aren't as many rigs heading up into the mountains with tipi poles lashed to the roof. Boulder was never that much of a "western" town. In the early seventies you might have seen a cowboy or logger roaming around once in a while, but they are hard to find now.

The only person I've seen with a cowboy hat so far was a Japanese "dude" who was looking over the Chinese cabbage at the grocery store. He was a bizarre, slicked up version of a 1950s black and white, shoot 'em up, TV cowboy. He was probably new to town and in his own way looking for the American West the same way I was when I rolled into Boulder for the first time.

Actually, Boulder today reminds me a lot of the Washington, D.C., metropolitan area that I grew up in during the 1950s and early 1960s. It's as fast and almost as slick. Everybody seems to be driving snazzy cars and talking about "doing lunch." I can't say that after ten years in smaller, slower places that it doesn't sometimes intoxicate me. It can be like watching some really exciting weather pour in over the mountains.

Forest Service district orientations for seasonals don't vary much from one place to another. On the Boulder district the emphasis is on dealing with the hundreds of thousands of visitors that will show up from the Denver area and from across the country. The Boulder ranger district may be the most "used" district in Colorado. I won't really be involved because I'll be working in the timber. You don't see many tourists out in the boonies.

The district ranger is concerned about how we deport ourselves to the public. We are charged to be "good hosts." The public is our customer. It's like we are running some sort of big hotel up there. She relates a story to us about an experience she had on another district.

The "other" district she is talking about was also a "high use" district. It seems like the off-road use of motor-cycles on the forest had come into question. The press wanted to do a story. She met with the reporter and every-thing was going smoothly until the reporter turned to a seasonal who was sitting nearby and asked him what he thought about the motorcycles.

"I hate the bastards. They're tearing everything up. Whenever I see one I wish I had my .30-06 so that I could blow them away," he said.

The ranger said that from that point on she was into damage control. The Forest Service managed to keep the quote out of the newspaper.

I quietly cheered that guy, not so much for his thoughts

on the motorcycle question, although I tend to agree with him, but simply for the fact that he had balls. He knew what he was saying. He probably also knew that it might mean his job on the district. But what do you really have to lose if you're just going to be laid off when the snow flies? That's the beauty of being seasonal, of being migrant, of being a poet. There's a high price to pay for sure, but if you are willing to go that light, there's the chance that now and then you'll find the perfect opportunity to say what's on your mind. A flock of geese flying south or a herd of elk moving back up country in the spring doesn't take our breath away for nothing.

I've stuck with it a lot longer than many of the other seasonals I've known. You'll see some who last only one season, then there's a group that may last three or four seasons. After that you run into what are called "long-term" seasonals. These are the people that for one reason or another choose to keep coming back year after year, season after season.

Many long-term seasonals have graduated with forestry degrees and keep coming back in hopes that they will someday land a permanent job with the Forest Service. In some cases the tenacity pays off and they finally get on board. In other cases they end up bitter and chuck it after eight or ten years, but it doesn't always last. After a year or two you will sometimes see them again on a different forest working for a new district. It gets into the blood.

For a while I thought I wanted to get on permanently with the Forest Service and jumped through some of the hoops and even lunged after the carrots that are often dangled out in front of the seasonals who are trying to come in out of the cold. I didn't have a forestry degree, which was already two strikes against me, but I tried anyway.

I began to realize how strange the relationship between the permanent employees and the seasonals really was. On one hand I think that many of the permanents had joined the Forest Service with the thought in mind that they would be outside all the time. Many of them even worked a number of years as seasonals before they came on board permanently. They found that they were spending more and more of their time inside as administrators, rather than outside as foresters. The seasonals were doing what they thought they'd be doing. A lot of them envied the kind of outdoor savvy you learn when you spend all your time "out there." On the other hand they could not understand why a person would devote ten, twelve, or fifteen years to a dead-end job with no hope of security, benefits, or a retirement plan.

The seasonals couldn't understand why a person would spend his life inside pushing papers in what appeared to be an endless succession of compromises, cop-outs, and concessions that couldn't do anything but lead to mediocrity, first on the job, and finally throughout a life. It wasn't that any of the people were any less strong than we were but more that the system they had devoted themselves to was so much stronger than any one man or woman. Most were bound to get ground up. We watched for the light to go out in some of their eyes and for the few very strong ones who were bound and determined to shine on.

It was nothing more than the age-old friction between dreamers and pragmatists. We all pay a price. When our bosses and supervisors are nicely retired, living in houses that they own, and with a few bucks in their pockets, a lot of us seasonals will probably be out behind the Safeway picking through the garbage and living in the back of our pickups.

I went into it looking for the line between the East Coast and Wild West. I needed to know where civilization stopped and the wildness began. I couldn't have helped it. I grew up watching a black and white Richard Boone, portraying Paladin in *Have Gun Will Travel* and reciting poetry as men were hanged. I saw *Wanted Dead or Alive, Gunsmoke, The Restless Gun, The Rebel, Wagon Train, The Rifleman.* . . . They were simple television melo-dramas where right was right and wrong was wrong. They depicted a time when a man could make it on his own,

if he could just get away from all the shit. I believed them
and figured the answers must be out west.

I wasn't alone. There was a mass migration of dreamers
to the Rocky Mountain West from the mid-1960s well into
the seventies. We came with our cowboy hats and boots,
rifles and bedrolls. Most of all we wanted to get away from
it, but we were also trying to *find* it. Boulder, Colorado,
was a good jumping off point. Our particular migration
wasn't much different than all the other migrations to the
American West.

The locals were actually kind of amused. The fact was
that they were, indeed, authentic and we were not. I ran
into people on Colorado's Front Range whose families had
opened the country up back in the mid and late 1800s.
Most of them were living comfortably in places like Denver.
Most of them said we didn't understand the West.

"You can't make a living back there in the mountains.
Hell, we never go up there except to ski," they said.

A lot of it was true. Their forebears had paid the price
of admission. There weren't many that wanted to slime
around in the mountains just to earn minimum wage and
get frostbite. But now and then if you looked hard you'd
come across an old, handscrabble son of a bitch who lived
off in some hidden little draw or behind a pile of rocks. It
was the image I was looking for. I wanted to hang with
the folks who knew the seasons. You can always find them.
They may not want to talk to you, but you can find them.

The frontier West that I was looking for doesn't exist
anymore. Sure, I've been in a backwater bar here and
there where somebody took a pot-shot at a friend or lover,
but that happens in Chicago or New York every day.
Things are pretty much regulated out here just like anywhere else. If you are doing something crazy it will sooner
or later come to the attention of the authorities. The
important thing is that if you aren't really hurting anyone,
it might be later. That is the advantage of the open
spaces. Things come to light at a slower speed. There are
remnants of wildness here. It is possible, still, to get

far enough away to at least experience an image of freedom.

A couple of seasons ago I was surveying some spruce near timberline. This is the best of jobs because you are out alone. It occurred to me then that I knew what I was doing. I had a feel for how the draws came together and where they might lead. I could figure where the trees might grow best. I could see where the fire might run if lightning nailed one of the old snags. There was a little rounded knoll that was thick in brush. It was the kind of place the elk would bed down in after a morning of feeding on the edges of the meadows. Farther down the valley there were some benches of aspen. This is where they'd go for the rut in the autumn and maybe return to in the spring when the calves were due.

Most of all I had a feel for every one of the trees I saw. One might have had it tough. Another might have had an easier run. Some were busted up. Some were sick. There were a few brutes. Now and then there was one that was making life miserable for his neighbors. They all had lives of their own. It wasn't a lot different than the town I lived in when I came out of the mountains on the weekends. I got the feeling we are all tied up in this together.

It's too easy to say. There isn't a question of life and death. We all accept that. It comes down to simple dignities.

I've kept at it because it has been too hard to think about a year where I didn't see the autumn or the winter or the spring or the summer come into the country step by step like an old friend coming down the road. I've probably overstayed my welcome at the Forest Service. I've cut trees for twelve seasons. Some of the time it has seemed like the right thing to do, other times it has been tough. It's a hard way to become an environmentalist.

It also has something to do with the other seasonals and the permanents who still have that edge in their eyes. The people that in a lot of cases I only know by a first name. The ones that come and go. The drifters and dreamers. The ones that you haven't seen for years, then out of the blue, on some fire, in the middle of nowhere, here they come down the fireline like some freak dust devil that blows in when your back is turned.

"Still at it?" they say and move on.

If you asked any one of these seasonals they'd tell you that they're no damned good and you'd best stay clear of them. But let the word get out that you're between a rock and a hard spot and there is the chance they'll be pounding on your door. It was a good excuse to go somewhere, they'll say.

So, here I am in Boulder, Colorado. It's where I first saw people who were looking west wondering what's over the next ridge. People who always had a sleeping bag rolled up next to the door. Even now, they still trickle in, sometimes in a beat-up car, maybe with a license plate from Ohio, Illinois, or Virginia. I can tell the ones that have it bad, the ones that would be seasonals.

My friend Bob Matson retired from the Forest Service last year. I think he'd been a forester for twenty-five or thirty years. He fought with them, now and then, all the way to the end.

"I may not have done as well in my career as I could have," he told me. "But damn it, I've seen the Mendenhall Glacier."

IN SEARCH OF THE SACRED

a pilgrimage to holy places

RICK JAROW

*This publication made possible with
the assistance of the Kern Foundation*

The Theosophical Publishing House
Wheaton, Ill. U.S.A.
Madras, India / London, England

© Copyright 1986, Rick Jarow.
A Quest original. First edition, 1986.

The Theosophical Publishing House
306 West Geneva Road
Wheaton, IL 60187

A publication of the Theosophical Publishing House, a department of the Theosophical Society in America.

Library of Congress Cataloging in Publication Data

Jarow, Rick.
  In search of the sacred.

  (A Quest book)
  1. Shrines.   2. Pilgrims and pilgrimages.   3. Voyages and travels.
4. Jarow, Rick.   I. Title.
BL580.J35   1986   291.3'5   86-40122
ISBN 0-8356-0613-9 (pbk.)

Printed in the United States of America

To my beloved parents

Acknowledgments

Sincere appreciation to my mentors and special thanks to Peter Rand for his encouragement and support.

Contents

Introduction:
Gates of Dharma

For the last few years I have been working with a remarkable Cuban healer named Orestes Valdes who is
deeply rooted in the earth and its healing ways. After
years of training in various Eastern and Western disciplines of meditation, it has been invigorating to discover
methods of working with energy which honor the soil as
well as the sky, namely the Shamanic and Native American traditions which Orestes is so deeply rooted in.
Through this ongoing discovery, I have continued to ob-
serve strong threads of connection and continuity of the
sacred teachings through all times and places.

Every so often in the middle of the day, when there is
a lull in the stream of people who come to see him, we
stop our work and cook up a pot of dark Spanish coffee,
sit around, relax, and tell stories. Orestes often speaks
about his own tradition, telling us about his childhood
and his training in the healing lineage of his father. At
these times, I am struck by how closely he captures the
essence of the concept of *dharma*, a concept, or should
one say an energy, which was and is a guiding force that
propelled me on my pilgrimage to the gates of *dharma*,
to the gateways of the sacred teachings.

Orestes understood that he was a healer when, at the
age of eight (or perhaps twelve as I have heard variant
recountings of his tale on different occasions), he fell into

1

Introduction

a river and almost drowned. As they were pumping out his stomach, his spirit guide, Catalina, appeared to him for the first time and told him that he would survive as he had important work to do in this life. "We all have our own river to God," he often says, slowly sipping the tiny cup of concentrated brew. "You've got to find your own river. If you swim in another's, you will drown."

The Indian word *dharma*, like the Chinese *Tao*, is essentially untranslatable. It has been spoken of as "religion," "sacred duty," "virtue," "cosmic order," and so on. Etymologically, it comes from the Sanskrit verbal root /*dhr*, which means "to hold," giving the sense of that which holds everything together. A saintly Indian scholar once described it to me as the inherent nature of a thing. The *dharma* of water is wetness. The *dharma* of honey is sweetness. But definition destroys, categories and recountings are of the past, and the past is fiction.

Inherent nature, on the other hand, flows through categories; its river is one of resonance, and this river, like the story, flows, never twice the same, yet always being forged, remembered, and retold. The remembrance that follows is a selected chronicle from a six-month pilgrimage to various centers of spiritual power, sacred places which are said to widen the scope and receptivity of the soul. And, surely, it must be this same river which passes through so many places, branching out into divergent streams. Seekers of the grail, Marco Polo and the journey to the East, pilgrims marching through Lourdes and Compostella all have had their tales.

But what is it, exactly, that is remembered and retold? And why tell an old tale again when the true pilgrims of our time are thought to be those who venture into outer space and uncover new microelectronic wizardries? Is it but a nostalgia for something forgotten, a worn-out romantic refusal of civilized advancement. Is it a search for some cosmic conclusion that can be called one's own? Perhaps it is an avoidance, a fool's effort to prolong ancient hope in face of the dismemberment of all things?

These are certainly real possibilities. But there is

another as well, an essence of balance, a mythical mode which the human mind needs in order to relate to the rest of its cosmic neighborhood. And it is the effort to attune oneself to this essence which propelled this journey to holy places, a journey which combines my own personal pilgrimage with that cultural pilgrimage which we all share.

My own pilgrimage began much like anybody else's in America. My early companions were neither classical volumes, nor any inherent sense of mission. I grew up with the Three Stooges and Soupy Sales, baseball, basketball, and rock 'n roll. Every sign I was given said that you had to learn to play the game and play it well. So I played earnestly and basketballed my way into Harvard University, quite an achievement I was told for a kid from New Utrecht High School in Brooklyn. Never mind my sustained feelings of unease or my sense of unreality about the whole thing. I was going places.

While I was playing this game, I would occasionally stop and think about certain questions like: Who am I? Where do I come from? What is the meaning of all this? But people just smiled and said, "You'll get over it," or "It's bourgeois malaise. You should go out and join the revolution."

And that is, in fact, why I went to Harvard, to join the revolution. I had already hung up my sneakers. The classrooms, buildings, and hallowed hall traditions seemed totally irrelevant compared to what was happening in the streets of Cambridge. For here was a new world, a world peopled with the most bewildering array of costumes, groups, and ideologies, all charging the air with an electric expectancy.

I remember the day I gave up on the world and became myself. I was sitting in a lecture hall at Harvard with 300 other students, listening to a Nobel laureate in biochemistry explain how the world had originated from a Big Bang and the subsequent combination of carbon, oxygen, hydrogen, and nitrogen. Taking the chance of exposing my stupidity, I raised my hand and asked, in front of the filled lecture hall, where this Big Bang came from. I received the sneering answer that such questions masqueraded for deep thinking but were, in fact, irrelevant.

Now I understand that the professor may have been right. After all, the Buddha in his benign way gave a similar answer to such questions, saying that they did not lead to edification. But at that moment, it dawned on me that, while few would admit it, nobody really knew anything.

I fell into an incessant seeking. I began to read successive books on psychology and religion. I walked the streets and talked to strangers who told me that Harvard Divinity School was filled with atheists and that a good deal of the graduating black seniors had gone back to Africa and then returned to the States thoroughly disappointed. My room became a crash pad for all kinds of street people. I sat with the radical groups out in Harvard square, the Process people with black robes and crosses around their necks, the Hare Krishnas, the Black Panthers, the Animal Liberation people, and so on. I opened my eyes for the first time and saw that everyone at school was lonely, that so-called learning was based on acquisition and fear. Both the haves and have-nots were walking around without acknowledging that essence, that very center, that shared heart which is so dreadfully absent from us all. I read more and more works of "literature." At least here one did not find the pretense to knowledge, save for those who professed the subject. Then one day my esteemed professor of World Literature said to me in a moment of honesty, "Go on, I don't know either and I'm not going to play father." That was the day I left.

I left my school. I left my home. I left my friends who were getting sick of me "French-frying" their brains, and walked out, into the world. "I hope you find what you're looking for," some said. "You'll be burned out in six months," said others. I became a New York City taxi driver in the days when the city was wide open, when you never knew whom you might meet in the back of your cab. I vowed to drop my conditioned distinctions and open myself up to all forms of experience. I was burned out in six months.

But I had saved up some money and, no matter how frazzled and disconnected I was, something had opened up inside, and there was no turning back. I decided to explore my own country and went hitch-hiking coast to coast. At that time, a very strange thing happened. No matter who picked me up—a black Baptist minister from South Carolina, a young American Buddhist on a forty day fast stopping at every gas station for another drink of water, a reincarnated Atlantean from the Edgar Cayce Institute—whoever it was would begin to talk about God.

This was something quite new for me. God, not as a theory nor as an organized religion, but as part of someone's personal experience. I wound up alone in a park in the Great Smoky Mountains of Virginia, and for the first time in my life I began to pray. I didn't pray for anything in particular. I just sat on the ground and prayed. I suddenly saw that my entire life was running headlong towards death, and that I had missed the most essential, the most central thing of all.

Something had touched me, had shaken me to the depths, and I understood that, if I could trust, my life would somehow be guided by a power that overswept everything I had ever thought of as my own. This wasn't any kind of philosophical understanding; it just began to happen, to intrude upon my days and nights whether I liked it or not, just like the intrusion of all the people of God who picked me up on the road.

At one point I was feverishly taking notes in the back of a pickup truck en route out west. When the truck went on its own way and I got off, I realized that my notes, my great, profound thoughts on life and love, were missing. I had left them in the truck. After my initial "all is lost" reaction, I said, "Okay, I trust, I accept it." Just as I said this in my mind, my foot struck an object by the side of the road next to a corn field in the middle of

5

nowhere. I picked it up. It was a volume of the sayings of Lao Tzu which spoke of the "Way of the Tao" as knowing less and doing less every day, as letting go into the constant, into the uncarved block.

I returned to New York and to my taxi, feeling that I had to let go of my entire past, of all the hooks and coercions of society, family, and friends. I had saved enough money to cross the ocean, and I became a pilgrim, a seeker of the way. I wandered through foreign countries, read books in European libraries, met people and stayed with them, worked in market places and on agrarian communes. I was introduced to various spiritual teachings, practiced yoga and meditation, sat at the feet of gurus, and eventually became an earnest devotee, adopting the role of a religious person. While in Europe, I also had my first encounter with cathedrals and monastic orders and at times took refuge there. For they seemed like long lost friends. The living encounter with other cultures opened my mind to a new sense of history as well, not that of the history books and their chronologies, but one in which a soul could look out and become aware of its awesome pilgrimage through space and time.

I was especially affected by the spiritual teachings of the East. Like most Americans, I had no practical discipline at all. I couldn't even begin to sit with my legs crossed or with my back straight and had been living on a diet of French bread and other assorted junk foods. I clearly understood that my life of T.V. watching and refrigerator raiding had made me a slave to my senses. How could you understand "truth," how could you go "within," if you couldn't even sit still and concentrate for ten minutes? The "spiritual life," when presented through Eastern ideals of meditation and right action, seemed like a more sane way to live. And there was a goal, a transcendental destination.

The concept of the spiritual teacher and his power was also an attractive one, for it involved much more than the power of a charismatic individual. Most teachers, even when they claim to possess a personal grace, derive their teachings and their power from a particular river, a lineage, a mode of transmission through which the sacred teachings are handed down. The lineage becomes the bedrock of the culture, the thread through which the rest of life's activities are connected. Eating, sleeping, working, and personal relationships are no longer seen as simply random activities. They are part of the fabric of the tradition, of one's *sadhana* or spiritual discipline.

I went on in this way for a number of years, eventually moving to India, practicing meditation, studying sacred texts, and living in the ashrams of my teachers. I developed a deep love for that land. I wore the garb of a wandering mendicant and visited the great *tirthas*, the holy sites of pilgrimage. Wherever I traveled, people were kind and supportive. The *sadhus*, nomadic aspirants, treated me as one of their own, feeding me and giving me shelter. Many, of course, were overwhelmingly curious, but beyond the fascination of a white-skinned *sadhu* there was a communicable goodness that emanated from these people. I could walk into a village, unknown and unable to speak the local dialect; all it took was one look into the eyes of a villager to make me feel at home and at one with the surroundings. I would be taken in and cared for by people who seemed to be among the most hospitable on the face of the earth. I fit in so easily in India, perhaps, a bit too easily. It was almost like being at summer camp (save for the occasional bouts of dysentery and malaria). And it was with a good deal of sadness that I slowly accepted the fact that I was not meant to stay and live out my life by Ganga side. For despite this seeming ease, my own culture continued to haunt me as it demanded to be repossessed, even if in its own special way.

I still often wonder why this simple life of the spirit wasn't enough for me. Why did I have to leave it all and find my own river? Perhaps I was hopelessly spoiled. Perhaps I had no real faith. Or perhaps I had caught that glimpse past innocence, the horror of the student who finds that the master is but the disciple's creation, another product of the clinging mind. In the end, one is alone.

7

Introduction

After seven years, I returned to New York and to my taxi. I lived in a loft with nine other people, all spiritual aspirants of some sort. The loft was divided into rooms by curtains, with a communal kitchen and washing machine. While we each had our individual spiritual practices, we all went to the meditation classes of Hilda Charlton, who had been on her own pilgrimage to India for eighteen years and was now our mentor. She held the energy for so many of us neo-Hindus in New York, with a remarkable combination of strength, humor, and love. We were in society but were not of it. Most of us held marginal jobs and made just enough money to survive. I was happy. I did my sadhana, meditated, worked, and even went back to school.

American sadhana had some unusual aspects which were lacking in India, such as going to movies and eating ice cream. And that was fine. We were evolving our own forms of spiritual practice, singing with guitars instead of organs or tamburas, wearing jeans instead of robes and dhotis, and putting Christ, Krishna, and Buddha on the same altar.

I got a job as a paraprofessional teacher, working with neurologically impaired and emotionally disturbed children. The school even allowed me to give them yoga lessons. But despite this, every day was more depressing. No one looked at the beauty of these kids or considered their practical potential. Instead we had to spend hour after hour trying to teach them to read and write. If they got out of line, we were to drag them off to the dean's office, by force if necessary (which was often the case). The entire classroom functioned around the mentality of "who's the boss." The students grudgingly went to school because they had to. Their lives at home were totally disconnected from their lives inside the school building. Every day they would be forced into their desks as frustrations mounted. As soon as school was out, they hit the street, releasing everything, just waiting to come of age and stay on the street for good.

I went back to my taxi, but New York was now a different city. Every cab was equipped with transparent bulletproof barriers to shield the driver. New drivers were warned about picking up blacks and going into "bad neighborhoods." But I didn't care. I kept my windows open, chanted my mantras, picked up anyone, and was held up at gunpoint next to Central Park.

The spiritual teachings say that you have to go beyond your physical body and intellectual mind. But my mind was curious and, worst of all, critical. I had swallowed various teachings and teachers whole. But while in India, I had discovered that the story was much deeper and more complex than had ever been presented to me. So I renewed my studies of literature and took up Sanskrit as well, wanting to explore the roots of sacred traditions for myself.

I still had to support myself and went to work selling peanuts in the streets where I was arrested for vending on an illegal corner. The streets were filled with struggle, theaters and their sex-vibes, newspapers, honking horns, and the rest. Everyone had to make a buck, and mendicant begging did not sit well in New York.

The main line of traditional spiritual teachings generally said that you should avoid women, avoid sexuality, stay high above the second chakra. But I had an affair with a dancer, a woman who understood the earth in a way that I never did. I found out that, while you may indeed suffer in a relationship, your spirit grows. I knew that I had been holding onto a mind set which no longer worked for me.

I had been taught to meditate for hours and receive "messages from on high." That is, I would sit and sit in the lotus position, and after coming down, instead of breaking the energy and returning to the world, I would begin to speak, not from a place of personality, but from a space which was unknown. Voices would come through. Energy would run through. My whole body would be filled with electric vibrations. I became acquainted with a whole new world of "higher beings," saints, Masters, and the like, and began to feel high all

the time. Minor miracles, like finding parking spaces in
New York City, seemed to manifest whenever they were
needed, and I half walked and half floated throughout
the day. Still, I wondered about those messages, miracles,
and the teachings that went with them. For I felt a sense
of lack, a missing solidity, and a lack of connection with
the earth and its people.

The peanut business was not very lucrative, and I
didn't particularly relish standing out all day in the middle of winter. So, despite being held up, I began to drive
my cab at night again. I would drive till three in the
morning and watch the city turn into a ghostly dream of
characters who looked like they had been cast up from
some kind of hell. As I drove, I wondered about faith
and my professed universality in face of the squalor and
grinding economic reality of the city. Was it all a mirage,
this airy devotion, this flight of the heart? Did one indeed
have to choose between God and the world? Or was it at
all possible to live in practical vision, in shared vision?

I felt that whatever spirituality was, it had to interact
constructively with the world, with the living, moving,
grinding machinery of time. Groups and teachers began
to seem like sorry substitutes for mommy and daddy,
crutches which one holds onto so as not to face a problem
which is, above all, one of cultural disintegration. I
wanted to enter life again, the mainstream, but not at
the expense of my soul. I had listened to the teachings of
society, to the teachings of the university, to the teachings
of the East, to the occultists, and the avant-garde, but I
still found myself without a center, without a true sense
of lineage, without a living metaphor with which to balance myself in a moving world. True, one is alone, but
there is this river with its amorphous fabric of traditions
and streams of lives, this current which one needs as one's
life blood.

So I set out on pilgrimage again, this time as a prelude
to accepting the world. I set out to understand my own
lineage, and to find a way of living which could create a
viable vision. I set out to revisit those who had helped me

along the path before, desiring to reclaim that which was my own and to see through that which wasn't.

There were some darker feelings which urged me on as well. While repelled by the melodramatic self-importance of the prophets of gloom and doom, I, too, had a growing fear that if I did not see these holy places, if I did not make the contact now, I might never get the chance again. I remember sitting in a zendo for a week in New York City, unable to detach my mind from the vibrations of hate which poured forth like exhaust fumes from the news media around the figure of the Ayatollah Khoumeini. Everywhere was this cancerous expansion of information, weapons, predictions, and I frankly began to wonder how much time was left on the cosmic clock.

Like a good American, I had no deep-rooted sense of the past, no tradition to call my own. Mount Rushmore just didn't seem awesome enough to elicit worship. The calls to the future struck me with the same counterfeit thud as the six o'clock news, and taking on the robes of another mythology could never, in my eyes, rid itself of pretense, of the infatuation with a new form unrooted in substance. No river can continue to flow without being connected to its sources. I needed to make contact with these sources, to see firsthand what they really held for me, to emblazon into my being those places where the holy presence had been felt, and to commune with centers of light which remained incarnate in the many shrines of the world.

I had no detailed plan as to which particular shrines to visit. Rather I would allow myself to gravitate to those areas which resonated with my own heart. Certainly there were others, many other holy persons and holy shrines which I would have liked to see. But if I could at least commune with these few special places, I would perhaps see my heritage evolve from within, a heritage which could be shared with my own community, along with an understanding of sacred geography, of the "holy

place," and the ability to appreciate the ground one
stands on, right here and now.

Pilgrimage, furthermore, when undertaken with the
right mental attitude, is more than an accumulation of
exotic sites. It is a means of purification, an enacting of
an internal process in the external world. The traveling,
the effort to arrive, the work to keep the mind and heart
open to the sacred manifestation, all this charges one's being with vision, with insight, and with purpose.

My goal was not to recapture something lost and trans-
pose it on quotidian reality, but to walk wide-eyed and
awake through the halls of the past, to abandon clinging
to useless forms while remaining receptive to the grace of
those souls who came to the aid of our own, to move into
my fate with a sense of living wisdom and humility.

The cathedrals of France mark the presence of that
Christic space, that glory which still filters through the
darkened naves, remnants of the medieval mind and its
reconciliation with the soul. The communities of farm
and forest mark the new culture which rises on the earth
in reverence and communion with the greater spectrum
of nature. The classical monuments speak of another
heritage, a river which has spanned eons, forgotten lands
which arise in memory, restoring ancient harmonies. In
India, holy nomads still walk through the many-armed
land, proclaiming a magical dance, a vision of wonder on
the face of extinction.

The pilgrim's way led to Jerusalem, to the wartorn city
of the Holy Land. Here is where the journey supposedly
ends, where it is done. The pilgrim's trail, the marching
of knights errant, the seeking of indulgences, and the
witnessing of miracles, all of this is done. Incarnation
brings on new forms, but the story remains to be shared,
while the river and its streams linger on.

The tale told here is a completion of a certain episode
and, perhaps, the prelude to a new and greater challenge.
When one ceases playing at being either the sagacious

wise man or the humble fool, one can no longer lead or
follow. All that is left to do is to share. And so, this story
is shared. It is shared in reverence for all who journey
and for all journeys which mingle with our own. And it
is shared for the community which rises, defying all
world categories as does the story itself, belonging to no
place, no time, and no person. Through the gates of
dharma many paths are trodden. In the way of the road
no answer is given. Stumbling road, error upon error. At
the depth of the valley, another mountain.

1
France

The journey began. I returned to the medieval spires of Europe, the roots of Christendom, to the chivalrous remnants of stone-walled castles, to the speeding modern cities, and to the slow-paced countryside where placid white cows graze down sloping fields.

At the outset I called upon the Brotherhood that throughout history has been called by many different names. Who they are, in fact, I did not know. Some spoke of the Masters of the Great White Lodge as those who in bygone ages had striven for truth and had attained, elevated beings who chose to remain in relationship with the earth, working from the inner planes to churn forth new ideals and visions for the race. Others saw them as archetypes, collective thought forms, which rose from the esoteric imagination of the culture. Whatever their origin, their presence could be drawn by the aspirations of seeking souls to act as guides along the path.

Exactly what this "path" might be was another question. The cows, grazing in their own tranquility, seemed to embody more of it than those who seek it. Nevertheless, if I attuned to this energy a luminous power would appear that would produce a sensation of rising from the normal locus of space and time, along with an awareness of a community whose threads seemed to stretch through eons of human endeavor.

15

With the mind essence calm and attuned, seed thoughts would unfold along this rising energy, creating a power which could manifest in diverse ways. Through the practice of attunement, I developed a deep trust in the reality and power of these universal forces. To begin with, then, there was faith that there was a "Way" and that its pursuit would lead one to be in the right place at the right time.

To walk in this way made me feel not only like a traveler or seeker, but like part of the movement of that One Intelligence, that life which seeks to know itself. When I looked clearly, wisdom was everywhere. But those were rare occasions, for I was still looking, like a child learning to walk while bearing his inner light.

The Land of the Little People

I met Bernard in Paris and within a few hours we were speeding down the autoroute as though we'd never been apart. Years ago, we had traveled throughout France together, selling posters, books, costume jewelry, and anything else we could get our hands on to help pay for the farm. The ashram itself had since moved from the farm to a large castle, where it drew the wrath of traditional Frenchmen, especially when the newspapers showed photos of the guru being driven up in a Rolls Royce. The farm, however, remained. It was now a hangout for "fringies," as they were called, those who were half in and half out of the ashram scene. They generally paid some sort of lip service to the lineage of gurus, but were determined to do things in their own way. When they had split off from the ashram and bought the farm, they had sent me an invitation to come and live there. "It is not logical that you are not with us—*ce n'est pas logique,*" Bernard said.

Before I could reply we were hit with two flat tires.

16

The Land of the Little People

Things were just as they were in the old days. Our jack wasn't working. As we tried to hail down a car, I wondered what the farm would be like and if Bernard and I would be able to communicate now that I no longer thought I had to eat brown rice and meditate at four in the morning in order to become enlightened.

When we got rolling again I tried to open the window, but it wouldn't move. I pushed harder until it jammed. Bernard leaned over softly and said, "Not like that my friend." He moved the window sideways, slanted it on a certain angle, and it slid down effortlessly. So, he was cool. In any case, Roland would be there. My dear friend Roland had never really fit in with any group as he was mostly concerned with his own visions. He was tolerated, however, because he was one of the first members of the ashram, and also because he had convinced many of its residents that he was some sort of wizard who could communicate with "the beyond." I remembered listening to his stories about his third eye opening up, how he would see cobwebs and devils around people, how gods and goddesses would visit him, and so on. I wondered what he would be into this time.

We crossed a wooden bridge and turned up a small dirt road and came to the gate. A wooden sign hung from a tree saying, "Les Grottes." We had to drive slowly as everything was receding into dusk. I saw Roland from a distance, and he spotted me as well, calling out, "Baba," with a deep voice. We embraced and broke into laughter.

Things at the farm had mellowed out over the years. Bernard had started out alone in the winter, rebuilding the inside of the stone house, making floors of cow dung and clay, and reading holy books. He had come through some hard times and was now on his own and staking his claim. Soon others came. They raked and hoed the land, planted cabbages, tomatoes, and cauliflower, and bought a horse. There was now a loose community in residence. People seemed to drift in, disappear, and then return

17

months later. But something was very much alive. Perhaps it was Bernard's own determination, or the land itself, but you had the feeling that whatever was happening here was going to last.

Folks sat around most of the night playing music and nodding back and forth. There were few introductions. It was just assumed that I was part of the scene. As the gathering wound down, Bernard and his wife returned to their own cabin leaving me in the hands of Roland, who sat down beside me and began narrating the details of his most recent life.

He now lived alone in a small forest cabin. He played the bagpipes and grew medicinal herbs for use in various teas and healing potions. Even though everyone had gone, Roland still spoke in whispers as he began to expound upon his favorite subject, the "little people" who dwelt in the woods.

"I've known of the angelic hierarchy since my childhood," he explained, "and since I've spent just about all of my life in rural areas, nature has always been a person to me; someone with whom I could share things, talk to, and learn from."

Now I knew that Roland hadn't grown up in the forest. In fact, he used to steal antiques from the Cligancourt Flea Market to make a living. But Roland always had a good story, and I was glad to hear him talking about angels. The last time I sat around listening to him, I had been told stories of a she-demon who had taken the form of an owl and had chirped in his ear all night in the Vallée des Mages in southern France.

As he continued to describe various types of nature spirits, however, I sensed a great change in my friend's demeanor. While always mysterious, Roland was much lighter than usual as he spoke of the particular functions of each type of spirit. Some created archetypal patterns for the world of plants. Others assisted in all sorts of cosmic functions such as the blooming of flowers in the spring, and others were in charge of the seasonal changes in the colors of leaves.

The Land of the Little People

As far as his own work with medicinal plants was concerned, Roland felt that it was very important to acknowledge the particular Deva or angelic spirit who was working with a plant or herb. "Before I pick *any* plant," he emphasized, "I always come and warn the Devas concerned a day ahead, explaining not only my reasons for picking them but just how it will be done. The plants I pick are then blessed with healing energy, and the results are truly amazing. The roots, stems, leaves, and flowers of each plant embody a particular aura, a healing radiance given out by one of the Devas or Nature Spirits. Different parts of the plant tend to work on different levels of consciousness. When we ingest them, a particular quality is enhanced within ourselves. Therefore, I plant as wide a range of herbs as possible, allowing the maximum amount of energy to pour through from the Devic realm."

We walked out through a field of tall grasses and onto a little path which led to the forest. The moon was almost full and lit the way where wild flowers and clusters of plants stood, looking like weeds to me. Roland began to explain the effects of these plants when mixed into infusions or balms. Pointing to a bunch of dark colored flowers he said, "Those will send you right off," then he paused and intoned seriously, "but you cannot be negligent or uncaring. If you are, the worst possible thing will happen. The Devas will make themselves and their world invisible to you. For all of the plant kingdom is personal, just like the animal kingdom, and certain herbs and flowers contain great power. Once you have been allowed into their realm, you must be extremely cautious. Why, do you realize that at any given moment the earth could just split open and swallow you whole? Who would know? You must, therefore, learn to discriminate, as there are benefic and malefic beings in the same forest."

We continued on the narrow lit path and Roland gave me further instructions. "Never do anything haphazard in the forest. Even if you just "go" under a tree, always ask permission first." He continued to point out various herbs

19

which were growing wild along the way. "By simply thinking on the little people, you are brought into contact with them, so be careful about the thought forms you are creating. They love good vibrations, anything that is expansive and free."

We reached Roland's cabin, which stood alone in a clearing about two hundred feet into the forest. The cabin was surrounded by beds of flowers and herbs. A set of bagpipes hung on the outer wall, which looked washed-out and pale from successive rains. Roland picked up the pipes and began to play while pivoting around on his right heel. He stopped, turned towards me, and said, "The little people are somehow attracted to Irish and Scottish folk music. They often come to hear me play the pipes." He then launched into a detailed description of various methods used to ward off the malefic spirits. Some of them were standard uses of prayers and mantras. Others were more intriguing, such as certain chalk formations which he kept on the cabin door and arrangements of selected herbs and flowers which he put by his bed. He also demonstrated how to ring a bell at timed intervals in order to induce receptive states in which the little people could be contacted.

We entered the cabin. Roland lit two oil lamps. The small room was very neat, and every article was in its precise place. Along the wall were shelves containing jars with labels, apparently different herbal mixtures. A bow and arrow along with a large target stood in the corner. Roland said that on the whole, his experience with the little people was positive as they did not generally misuse their powers. Nevertheless, one had to take precautions as the little people were generally suspicious of human beings and would always test your intentions as well as your capabilities. He invited me to spend the night with him in the forest and then excused himself saying, with a mysterious voice, that it was midnight and he had to say hello to a few friends.

We spent the night over a crackling fire. Sparks flared

upward and disappeared into the moonlit smoke. Roland told me that he had been back to the Valley Des Mages, this time accompanied by a Peruvian medicine man who had him drinking quarts of olive oil with lemon juice and then fasting for days until he saw visions.

As the night passed, we began to reminisce. "When Swami first came to the ashram," Roland explained, "he quickly saw that the Western disciples were too self-motivated and frazzled to understand meditation truly. So he gave them what they wanted, achievement and success. They could collect money and live in castles. But this is not the way of the wizard."

Roland's eyes danced over the fire. His head seemed to expand with the moon. "So now I make tea. In a while, I may be doing something else." He mixed up a batch and put it in a kettle, lowering it over the fire. "For me, mixing herbs is like making love," he said. He had brought a few jars out of the cabin. He took small amounts from each jar, considered them carefully, and added them to the brew. "Remember," he repeated, "when you pick herbs always ask permission first."

I cannot say exactly what the tea did to me, but I began to feel a light, spinning sensation. I became transfixed in the fire, then on the wind flowing through the forest, and then on Roland's voice as he spoke of the mysteries of nature. I do not remember much, except that he elaborated upon a hierarchy that was to be found in every forest. There was always a king and queen tree, and they had to be sought out and propitiated before one could safely reside in a wooded region. "We must be respectful of their realm," he intoned, "especially during the changing of the guard." It seemed that around dawn and twilight, different nature spirits came "on duty." "You too can be a wizard," Roland said. "Anyone can commune with the natural forces. It is only confidence that is lacking."

I felt myself dropping off to sleep, filled with the warmth of the fire and the mixture of forest herbs. I

21

nestled into my sleeping bag, rolled over the leaves on the ground, and smelled the earth and fire. Something in Roland's new world spoke to me; the jars of collected herbs, the bagpipes and arrows, the buckets for collecting rainwater. All these appeared as tokens of a new kind of life, a life which I felt in the forest sounds, in the red and yellow scarf which Roland kept wrapped around his neck, and above all, in the feeling of friendship with the woods and their elemental inhabitants. I noticed that Roland had himself taken on certain elfish characteristics; a spry gait, a certain gleam in the eye, and an extraordinary alternation between serious and clownlike moods.

We rose with the sun. The dew was coating the greenery of the forest. I looked around at the trees, their leaves illumined by the first shafts of daylight, and fancied that I saw tiny fairies curled up within each leaf. I gazed upon them as they lay serene and peaceful in their leafy hammocks, and one of them appeared to rise and smile at me in recognition. The moon still shone in the sky, waxing from Cancer into Leo, bringing an influx of new energy. I walked around to the back of the cabin and took a bath with rainwater collected in buckets by an ingenious gutter system constructed along the cabin roof.

We both sat, watching the sun rise. Roland took the black iron kettle, washed it out, and began making another infusion as he revived the night's fire. I told Roland that during the night I dreamt that my beard was growing. "That is good, Baba," he said, "because last night I saw you with a long, fiery, orange beard." He laughed and took up his pipes. "By the way," he asked, "did you see any of my little friends this morning?" More laughter. The pipes sounded through the trees and danced around the morning songs of forest birds. The water bubbled in the cauldron and the smell of fire came all around.

In meditation now, the sun risen over fields and forests. Breathing in the breath of life, the pure life energy of the real—earth, fire, and sky; the flower opens slowly. The fruits of your journey shall unfold in single steps. What has lain dormant now moves into manifestation.

While change brings tribulation, the light shall surely have its way.

It was time. I said goodbye to my friend, who did a little Irish jig and walked me down the path to the edge of the forest. I knew that we would meet again, perhaps clothed in another vision. But now, there was the road, and I was on it. I stuck out my thumb and caught a ride down south.

The Earth

Once, at a rock festival in Tours, a young English man wearing yellow tinted glasses and a bowler hat had explained the intricacies of hitch-hiking to me. "The hitching pose," he said, "is a special asana [yoga posture]. While waiting in your position, you must be absolutely attentive to your physical and mental being." He balanced himself on his left foot and deftly stuck his right thumb in the air. "Each passing car is one identical substance cased in a different form." He removed his hat and waved it around the open field. "You enter into the vibrational field of a particular vehicle until the scene dissolves and leaves you back on the road again." He made a sweeping movement with his arm.

Since then, I had cultivated the pose quite frequently. I was not too anxious about cars stopping or not. Often, a long wait would bring a ride with a most special soul. Besides, there was no better way to get the pulse of an area than to be out on the open road.

That morning, after leaving the farm, I caught a ride with a gentleman who raved on and on about the recent socialist victory in the national elections and how the United States was going to destroy the French economy by increasing the value of the dollar. I was most impressed by the French pronunciation of the names of American presidents, "Car-teur, Re-gan," always with a slight touch of disdain.

We passed by a series of enormous grey structures

which jut out in a desert area along the autoroute. They looked like inverted funnels and stood alone, surrounded by barbed wire fences. My driver explained that these were recently built nuclear reactors, part of the government's massive power program. Looking at those lifeless towers, my body began to shudder until I didn't want to think anymore. I just watched them fade, their immovable shadows cast long on the sand. There was not a cloud in the sky.

I bathed in the ocean by Juan-Les-Pins, walked through town, and traveled a few more miles up a winding road. There I came to a group of old houses with some land in the back containing rows of plants and a large greenhouse. When Orissa opened the door, she gasped in surprise. It had been five years now, and I was unannounced and unexpected. Her face was clear and mellow, and her brown hair was now only down to her shoulders giving her an easy, domestic air. It looked like her stomach was again beginning to bulge under her faded red shirt, and her little elf-child Mallaki was strutting through the house. John arrived in the early evening. His face was textured and sensitive. His eyes were still squinting from the sun, and his black hair fell over to one side. Without bothering to change, he picked up his guitar and began to play music of the earth and sky.

The next morning we went up into the mountains. John and Orissa had found the traditional practice of yoga too confining and were gradually moving towards their own inspired dance of movement and sound. With the sun coming through the trees we wound our way up the mountain. Earth mountain was the ever-present female, the World Goddess with her deep energy abounding everywhere—in the soil and stones, in the wild flowers and forest vines, in the soft, quiet ponds, and in the rushing falls.

We stopped by a natural pool formed by a circle of rocks which dammed up the falls. The pines, reaching upwards, seemed to grow out of the rocks themselves,

firm in their stillness and elevation. The water washed
over the rocks and its rushing sound mingled with the air.

We took off our clothes and slipped into the water.
Orissa stood with her arms and hands extended outwards
as water glistened into beads on her body. She softly
chanted into the woods, establishing peace with the
denizens of the region. Not moving against the grain of
life, the spirit in form becomes the form of spirit. In that
communion we were purely alive on the blissful body of
Mother Earth.

That evening, John spoke of the pact he had made
with the earth, to work with her and learn from her. He
had plans for farming with neither plow nor fertilizer, a
technique which he explained had been used in the Far
East. He also spoke of organic and biodynamic farming in
Ardeche, which I knew little about. So I listened for most
of the evening as he discussed the various earth communities and played a tape of Lanza del Vasto speaking of his
own community in the southern Alps, L'Arche, which
was founded on the ideals of non-violence and self-sufficiency. The community members wove their own cloth,
farmed, and collected drinking water from fresh mountain streams. Del Vasco had spent time with Gandhi on
his own pilgrimage to India, and had returned to begin
this work.

I remembered my one visit to his mountain abode. We
only spent a week-end there, but it was like visiting
another planet. People wore blue hand-sewn garments,
were expert in all sorts of handicrafts, and gathered
together every evening with one of the community
"elders." There were a number of Arche communities in
the mountains. Each one had its own particular feeling
and lifestyle. Yet they were all connected, energized by
the pioneer spirit of del Vasto himself, and led by one of
the elders who acted as a focal point, mediator, and
spiritual guide. My most vivid impression of L'Arche,
however, occurred on the morning of our departure. Our
van would not start, and within two minutes more than

ten people stopped their work in the fields and came over to give us a hand and help us get moving. That said it all.

As we sat through the evening, surrounded by the sounds of earth's creatures as constant affirmations, I thought of my own ideals. I wanted to discover life's healing forces, to awaken my deadened senses, to recover my awareness of those forgotten beings; the creatures who were hiding away from man in fear, the Devas of the fields and forests, and the angels of the upper regions. Above all, I wanted to rekindle that which I was meant to be and that which Earth was meant to be. We sat on the porch and formed a healing circle, softly intoning vibrations of harmony which expanded outwards into the air.

Notre-Dame-de-la-Garde

A few days later I hitched down to Aix-En-Provence and ran into Jojo, an old hanging out-friend from Paris, right on the Cours Mirabeau. He was leaning against a column on the street smoking a cigarette, as if he had been waiting for me. Although we hadn't met for many years, he didn't act the least bit surprised to see me, and he took me over to Patrick's house which had been my destination. I had last seen Patrick in a cottage by the Ganges hanging out with the ganja-smoking Shiva Babas and carving stringed instruments from gourds, going by the name of Prabhavananada-ji, or something like that.

He wasn't there when we arrived, so I spent the morning washing out my hitching clothes and drying them in the sun of Provence. The air was arid and fresh and mingled with the scent of herbs and all sorts of sounds; the wind, passing car radios, forks clanking on plates. I wondered what it would be like to hear purely, with no mind filter, a wonder that was enough to break the energy of silence.

Patrick and his woman from the islands, Angela, cooked an Indian dinner. Patrick was a Zen Master at rolling chappatis, Indian flatbreads. He worked with one hand, kneading the dough in a copper pan, slapping it down, and flattening it with deliberate movements. Not making a slip, he rolled them around and picked them up in one sweeping motion. Then he lightly held them over the fire, turning them until they puffed up soft, fluffy, and thick. Placing them one on top of the other, he wrapped them in a cloth to keep warm. That was one good thing about living in India, you learned practical skills like cooking with raw ingredients, cleaning pots with the husks of coconut shells, and cleaning yourself with water instead of toilet paper.

We had not seen one another for quite a while, so we passed the evening in nostalgic "remember whens," especially when Sahas came in. Sahas, passing through as always, had been abandoned in a hotel room and left to fend for himself at the age of eight. But, being a child of God, he always found a way; selling magazines for the Poor Sisters, selling records door to door. Sahas could sell anything. His latest scheme involved traveling cross-country in a fourteen-year-old Peugot selling silk-screen posters from Hong Kong at a hundred percent mark-up. He was so good that in a solid week's work he usually made enough for the next six, all in undeclared cash of course. He would then return to Aix, where Patrick had given him a room, and tend his holy shrine. He had a room overlooking the garden in the courtyard, and every morning he would pick choice flowers to garland the Shiva Lingam which stood in the center of his shrine. Sahas had all sorts of paraphernalia for worship surrounding the Lingam, lamps, bells, pictures, plates of sliced fruits, and so on, as well as a new television set sitting in the other corner of the room. All of these came from his marketing skills. "Eight and eight, that makes ten *chez-nous*, madame," he would bark as customers passed by, "because I don't know how to count."

Sahas had been in our ashram for a while and had

traveled with Bernard and myself, giving us an education in marketing and salesmanship. At the time, he was going through his holy self-righteous stage, characteristic of new converts to any system from religions to how to buy stocks. We had been invited by some Buddhist brothers to spend the night at their retreat in Toulon-son-Arroux. The atmosphere was very quiet and holy as a Lama had just arrived from Tibet to conduct a week long meditation intensive. While Hinduism theoretically agrees with Buddhism in its compassion for all forms of life, high-caste Hindu culture makes sharp distinctions between various living species in its effort to maintain "brahminical purity." Therefore, the next morning as we were calmly having breakfast with the Lama and his disciples, Sahas suddenly jumped up on the table and began castigating the Lama for allowing a cat to jump up and nibble off his plate. "This level of behavior is not on the level of human beings," he yelled, taking on the airs of a Hindu holy man. The Lama did not appear to be the least bit disturbed, but everyone else at the table blew up, and breakfast ended in an uproar.

Sahas, a good deal mellowed out since then, now had his holy shrine in one corner of the room and this T.V. set in the other, apparently alternating between them. He came over to me later in the evening and with a wry smile said that he wanted to give me something. He took out a handmade silver amulet inscribed in Sanskrit with the holy names of God and fastened it around my neck. I wouldn't dare ask where he got it from. I then told him that I would like to give him something and took out a packet of *vibhuti*, holy ash blessed by Satya Sai Baba, the Indian holy man. I explained how this ash, due to its spiritual blessing, had curative powers. Without any thought or hesitation, Sahas took some of the ash on his fingertips and applied it to his right leg, which he said had been badly bruised the day before. The next morning Sahas was singing as he fixed chicory root for breakfast. He showed me his leg. "It's cured! *C'est puissant, ton truc,*" he said.

28

Notre-Dame-de-la-Garde

After driving me around Aix and showing me its wonders, Sahas took me to Marseille, passing the Valley Des Mages on the way, that mystical abode where seekers and occultists of all kinds have spent many nights trying to fathom its wonders. Roland had always insisted that if you remained there for even one full night "something was bound to happen." Before going off to work the large Arab market, Sahas dropped me off by the port and asked me where I was off to. I pointed up towards the mountain where the statue of Mary could be seen overlooking the city. "*C'est bien,*" he nodded in a contented manner, "bien."

I arrived at the foot of the hill. How I remembered this place where I used to pray, and Mary's golden form rising high above the city on top of the mountain cathedral Notre-Dame de la Garde, Our Lady of Protection, overlooking the sea. The way up was a narrow, winding road intersected by flights of stairs. At every landing I felt her presence coming closer until Our Lady appeared directly above.

There is no answer, no way to describe the experience of Mary, of who she is or who she may be. She is, of course, the Great Mother in another aspect, the wonderful flow of grace, the aura of luminosity which covers the entire mountain where pilgrims have made their way since early morning. High above the teeming city, golden Mary holds the beloved Christ child in her arms. The child's arms are raised in the air. Under Mary's golden crown is a look of profound serenity. Her intercession, beyond measure, gently brings one back onto the path as she protects sailing vessels, keeping them on course. A huge crucifix has been placed over the concrete railing by the edge of the mountain. Walking over to it, you may gaze far over the Mediterranean Sea, out into Africa. There, one feels a slight change, a subtle scent blowing over the water. There lies the origin, a presence awakened in memory.

France

The cathedral was filled with model ships confirming the Mother's protection. Now there were only thoughts of Mary, the Divine Mother, the archetype of all mothers. All are her children in the purity of a mother's love for her child, in the simplicity of the heart. She is approached with the call of the child for its mother, not having the slightest doubt that she will respond. Her golden form watches over the city and its sagging ports. Her compassion extends ever outward; by the beaches, the drug haunts, and burlap market places. She is there.

I entered into the quietness of the great cathedral hall and slowly walked up the rows to the altar and knelt down. I had no rosary beads, so I said the rosary on my wooden Indian neck beads. There were others present in the cathedral, but as my eyes fixed upon the image of Mary by the altar, everything else disappeared. With Mary, there came a feeling of total dependency and refuge. In her presence, the child of the soul appeared. In the abundance of her grace, I knew that I would be guided and led somehow. I was not brought up with much religious training and did not know who Mary was or what she was supposed to represent, but the child within me knew and called:

> Hail Mary Full of Grace.
> The Lord is with Thee.
> Blessed art Thou among women,
> And Blessed is the fruit of thy womb, Jesus.
> Holy Mary, mother of God
> Pray for your children now
> And at the hour of our attainment.

I knew then that I was the Mother's child, and that her love was with me, no matter what I appeared to do or be in the external world. I knew that her love could be neither bought nor coaxed through any form of religiosity. It was already mine as her eternal child, no matter how far I might have wandered or ever would wander.

Notre-Dame-de-la-Garde

This, the child knew, and when this child appeared, the crust of my old personhood would fall away.

Later on, after walking around the darkened cathedral, I stopped by the far end to look over the various publications stacked on a rack by the door. While leafing through a magazine, I noticed a presence about and saw a black robed priest standing before me. He motioned to me to follow him, and we passed into a small cubicle of an office by the left side of the main chapel. He had grey-white hair and moved with some difficulty. His face was dry, but his moist eyes held you with a certain sadness. He asked me about my visit, and as I spoke, he opened one of his desk drawers with some difficulty as his hands were shaking. He paused, and then took out a set of black rosary beads with a silver cross at the end. He placed them in my hands and said, "For you, pray for me on your pilgrimage." I took the beads and thanked him, trying not to stare at his hands.

As I walked down the mountain, I tried to handle the beads with gratitude. I wanted to take them as a sign that Mary was with me, but I kept seeing those hands. Briefly, so briefly, our hands had met. But we could not stay together, so the sign of the cross was put between us. I began to hum a song to the Divine Mother which I had heard:

> I looked for her in the garden
> and She was standing there
> high up on a mountain.
> She's white as snow.
> She's the image of your soul
> Angel in gold. . . .

The streets were deserted. It was late afternoon. The desolate expanse stretched from the sweltering grime of the ports to the oil slicked beaches. From here, Our Lady seemed far away. Still, I knew she would always be waiting. Those hands and veins—I tried to move my fingers but the backs of my hands were filled with pain. I looked once more towards the Neptune-sea. Tides tugged on the waters of feeling, tides which my soul felt and trusted more than I could ever know.

Paris Revisited

As a point where experience converged, Paris, the center of the mandala, presented that reconciliation which could deeply open one unto Light. The shadowed confluence of memory walked with me down the Boulevard St. Germain, meeting a hollow-eyed, bearded traveler who said intently, "I am going no-place in the middle of no-where," and recalling the consuming sadness of trying to keep up with friends who picked pastries along the Rue de Rivoli, that sadness which appears when you know that something is dying.

The reek of tobacco hung over the streets like fog, and cafe smells mixed with new smoke already floated through the early morning air by St. Michel. The lamps had gone off, and under the lampposts the bookstalls were still covered over. To move through all of this without destination, to see through all of this without wanting, was true courage.

Approaching the bridge, I flashed back on the pain of the seeker, the "Fool" of the Tarot who looked ahead at the kaleidoscope of possibility, fascinated by the images of the known world, and who then saw, suddenly saw the expanse, that greatness beyond the abysmal, beyond the shady lunar water of dreams. Not yet equipped to name and thus destroy, the Fool sipped from their conjuring bowl until the waters drained into bitterness.

I had been sitting at a cafe table near the Bibliotheque Saint-Genevieve with the same woman I had been sitting with for three weeks. We had played out the personality game, the literati game, the love-making game, and were getting sick of each other. She told me how Frenchmen didn't like fore-play, and I told her about ashrams. Finally she said in disgust, "If all you do is talk about these ashrams why don't you go and join one?" So I did, vowing that I would never waste another minute of an incarnation at a cafe table.

Clearly, I didn't belong here, the city of sin some called it, and yet it was the sin that first drew me, the sin and the smoke, the cafes and the clubs, all the women you might meet, all perfect faces, all the libraries and sophisticated discussion on "culture and art." It drew you into the whirlpool, over the line, but through it was the seeking, through every eye that caught your own, the seeking, the same seeking. Standing on the bridge, flashing forwards for just a moment, I saw through time and through the seeker. Seeking, the very cause of unrest, had to exist if only to dissolve itself. Through the massive impersonality of morning traffic, a bridge crossed the Seine, the bridge where seeking may become acceptance.

There was a fogged stillness over Notre-Dame. The blood-red sun strove to break through the sky like the flaming sword of Saint Michael. The bridge crossing was a magical doorway which took one back in time, the bridge of past and future crossing the movements of the tinted river.

The carved iron doors of the cathedral were closed, but the blond Gothic stone exuded a depth which merged with strength as the foundation. Invisible threads of luminosity extended upwards from the spires as a continual pulsating aura of prayer. In its form and shadow the cathedral of Notre-Dame breathed a depth, a depth which warmed the body, breaking the hold of the outer play of the world, opening subtle senses which were delicate like flowers, a depth which turned in towards the foundation and crystallized even further beyond that, at the source. The mandala of experience led to the center and the center turned inward on itself, returning afresh, awakened, and anew.

I walked out in the sun a few blocks over to the
Georges Pompidou Cultural Center, the current cultural
monument which looked like the inside of a car engine. A
constant crowd hung out in the front area facing the glass
walls with their criss-crossing tube escalators. Many were
still sleeping, lying on the ground. Others were sitting
and passing around bottles of wine. The atmosphere was
similar to the summer festivals at Avignon; filled with
white-faced mimes, jugglers, fire eaters, rock bands,
fakirs who danced barefoot on broken glass, soap box ora-
tors, and the like. Crowds formed and unformed around
the various acts which went on simultaneously like a
series of side-shows. Further away, the cafes, custard
stands, and souvenir shops overflowed.

A large crowd gathered around a group of musicians
playing loud, syncopated music on two sets of conga
drums. Figures in black with painted faces moved around
them. The crowd grew larger and the music louder with
hangers-on, hustlers, and the curious all moving to the
drum rhythms. The vibrations rose in the air. The fallen
angels played in hell. The drums generated their fever;
louder sounds, brighter colors, canned laughter, decibel
ladders stretched across the void.

Later, I moved on alone to a quiet place across the
river, sat back in a small cafe, and looked up into the air.
Clouds moved in formations. Hosts of bodies moved
through the streets. If one didn't freeze them with the
mind, there would just be movement. My vow to keep
out of cafes was another partial imitation, another pose
limiting the expanse. Freedom had its own movement.
The Tao asked nothing and sought neither men nor
angels. Anywhere, anytime, the doors could swing open.
To be in a hurry was the disease of the mind.

Montmartre

It was a long climb from the base of the mountain to the
cathedral. Montmartre, the ancient "Mount of Mars," site

of siege after siege, was baptized in the blood of the martyrs. Now, a series of shops lined the base of the hill.
African immigrants with spread-out blankets sold carved
wooden statues all the way up the winding stairway.
From above you could see the throngs of Barbes, the
metro, and the funiculair or cable railway, and further
up on high, overlooking the entire scene, were the
whitened domes of Sacre-Coeur.

I walked inside the cathedral and sat down just as the
Mass was about to end. Two priests stood straight like
statues by the altar with cups in their hands and passed
out thin wafers of communion to the lines formed in
front of them, intoning *"le Corps du Christ."* Dreamlike
organ tones swept through the stilled air. The church,
lulled in the quietude of reverie, became ethereal, its
solidity transmuted into another substance. The mon-
strance holding the consecrated Host stood on high,
bathed in dazzling light. I shuddered, taken aback. Then
I rose and crossed myself as if doing so for the thousand
thousandth time. The music rose and wove like vapor
around the stations of the cross and up towards the mon-
strance which was suspended in its radiant presence, well
above the circling movements of time. I stared, enrapt,
and heard the music resonating like waves, rolling deeper
and deeper within. Its echoes were those of the Spirit,
asking the heart to open to that secret place where the
true being acknowledges its true need. How deeply I felt
the need for God at that moment. In this feeling there
was acceptance of subordination to that way which could
never be understood. In this feeling I was no longer my
own, and with it came the strength to follow this other
kind of knowing to the very end.

I remained in Paris for a few days, roaming the back
streets and retracing my steps through Cluny, St. Sulpice,
and further on. These were my last days in Lutetia, the
city of light. From the rivers and forts to the cathedral
spires winding upwards out of mass misery, the past hung
heavy over the city like the stale air itself. The mounts

and their ancient battlements, the romantic gardens of
courts and nobles, the struggles of the commune, the tri-
umph of factories and heavy industry, the cynicism, modern disgruntledness, and despair, the lost wars and
vanished power, all of this haunted the streets and more.

Autumn was approaching and soon would come winter
and death. What would become of the old culture?
Would the European effort, the patchwork of politics and
Christianity, be blown away with the turning leaves of
autumn, or would some monuments remain as a bedrock
of the new? Would the nuclear reactors transform all into
a mutant horror? Would the cities collapse and fall, returning to the primitive?

Along the right bank, bordering the Seine, was a tree-
shaded path running parallel to the highway. Under one
of the stone bridges was a small entrance into the concrete walls which led under the streets of the city. I
entered and poked around inside thinking that I might
get a look at the famous Parisian sewer system. I passed
through a rusty iron grating and followed a dim corridor
for about thirty feet as the light retreated behind me.

There, by an intersecting passageway, was an open
room with a few mattresses scattered on an earthen floor.
Pieces of clothing lay strewn about, and various belongings were hanging from nails driven into solid walls.
Sickly smells of garbage and alcohol reeked through the
cavern. Two men and a woman lay at the far end of the
room on makeshift beds. They were elderly and sickly
looking. Their faces, staring me down, were hardened
and plastered tough. Startled and annoyed at my entrance, a woman with stringy, dirty grey hair yelled out
gruffly, "What do you want? Get out of here!" I assured
her that I was friendly and had just stopped by to chat,
but the woman had already picked up a stick by the wall
and was pointing it menacingly at me. "*Allez-vous
en!*"(Get the hell out of here) she began to scream.

"Okay, okay, I'm going," I said, but I still couldn't

take my eyes off this underground world. Hundreds must be living down here, surviving off of collections pulled from cafe garbage heaps. Their possessions were in these caverns; some strips of cloth, empty bottles, mementoes from the past. I turned back and walked out into the busy daylight.

As I walked past the Louvre, I noticed a woman whose arms were filled with bundles, asking questions in broken French to an indifferent police officer. She was making desperate gestures, trying to explain that she was lost and had no idea of where she was going. I walked by, still numbed from the underground. I knew that I was missing an opportunity, a small opening into life, but I was tied to my own burdens. Should I or shouldn't I? It was too late. Busy human beings scurried in and out of department stores. I watched and wondered how much longer I would be busy, missing the opening in the name of some goal, or imprisoned in the loveless pose of the expert. I walked through the streets of the city and wondered if release would ever come from the biting envy of direction with which I guarded and possessed my life.

I was still walking by sundown. How to be alert, to meet the moment when it appeared? Climbing a stone stairway, I crossed back over the bridge from Île de la Cité. The sun was setting over Notre-Dame. It passed through the flying buttresses leaving shadows on the small green shops along the quay. It bent through the archways and cut along the bridge lighting up half the river.

I would have liked to linger in the dreamy mandala, in the darkened backstreets with their market place smells, or by the old bookstores launching new journeys. But all memory was passing, like the lights on the moving waters. Waters of the evening, stupors of every kind, processions of cafe meetings, people and places, pieces of life strung together and dissolving. Tomorrow I would move northward to the clear Flemish air, to the cities of the saints, to seek their inspiration. I said good-bye to Paris, city of women with flowers, where beauty became monstrous in her lack of form.

France

The Cathedral of Amiens

The train lines formed a network which spanned the
northern cities and spread out through Brittany and on to
the west of France. I had purchased a three week Eurail
pass which allowed you to get on and off the trains at
any place and time for no extra fare. I would travel to
one city, visit a particular shrine, and then hop back on
the train and go on. Sometimes I spoke with people on
the train, but mostly, I stayed inside myself, cultivating
the feeling of the saints which could still be perceived
throughout the land, even amidst the raging groans of
war and dying soldiers which still permeated the air, especially around the northern churches and cathedrals near
the Belgian border.

The old villages were rather symmetrical with the cathedral positioned in the center, often accompanied by a
cemetery. The air was still thick with gasping cries of
war. Perhaps there were many who were still fighting on
some astral plane, who hadn't acknowledged their own
death and were still caught in the drama of battle.

As I passed through one village after another and visited one shrine after another, I became increasingly
aware of the shadows of former wars; schisms, invasions,
inquisitions and the like. All were still present. The air
was still pock-marked and seething beneath the apparently placid surface of northern France and Brittany.

So I sought out the saints who shone through all this
like pristine stars, standing apart and shedding light, a
light which I hoped could somehow guide me through the
current debacle of war, terrorism, and power politics—
the world in which I lived. In all this time, nothing had
really changed.

After visiting Dunkirk and Calais, I turned slightly
southward to Amiens, home of one of the great cathedrals
of Europe. Blocking out my anxiety over what appeared
to be inevitable, I tried to enter into prayer. One had to
feel humble standing by this edifice, this holy ground.
The shrine was more than a symbol of the "inner sanc-

tuary." Its own presence could resonate with the aspiring heart and move beyond the clever and cutting cynicism which remains veiled in its own judgments. The mind itself must open, must set loose the spirit from its history of apprehension. The opening appears in the glimmer of humility which is the preparation for this willing death.

The nave exuded awe, deepening into silence, into the premonition of another world. This light, this truth, this cathedral, lay within, and yet the symbols and icons, sanctuaries and spires were created by men. Men, molding stone with their own hands, had turned the hard stuff of earth into this awesome vision. Sunlight sifted through the stained glass and gleamed along the organ pipes, spreading into a fan of rays about the nave. In this meeting of light in darkness man becomes vision, not in time, but in the moment of acceptance, of turning towards the call of the heart.

I wanted so much to remain in tune with the sculpted walls and soaring spires which set the mind's aspirations on high, but how was one to live, to really live without losing the vision, even while among the napkins, spoons, and coffee grounds?

I sat back in a cafe by the cathedral. Visitors paraded in time. Like rings circling from a stone thrown in a pond, all beings were swept in the movements of nature. And yet, there was a center from which all this radiated, and somehow the courage had to be born to build cathedrals from the very stuff of movement and flux, that living itself would become the doorway into the inner worlds of beauty and light.

The Saints

Captains of cafe tables, psychoanalytic theoreticians, technocrats of all kinds, and even musicians of silence had relegated the saints to the realm of popular fantasy and wish fulfillment. And this was in one sense undoubtedly true: a saint for every day of the week, a saint to cure headaches, a saint to invoke in case of lawsuits, to protect asses, and so on.

But old habits die hard. The altars of the Blessed Mother still dotted the countryside, surrounded by flower offerings and various propitiations. Saint Anthony, Saint Francis, and others remained on the walls of homes, or on neck medallions and key chains. They turned up on car dashboards and in store windows, and why not? After all, they brought good luck. And there certainly was some strange, unknown mechanism at work, as in the time when we prayed to Saint Ignatius to start the ignition on our van. The battery was supposedly dead, but it started right up in zero degree weather. Yes, one still heard of "little miracles" and new apparitions of the Virgin still appeared from time to time, causing a stir in the local journals and inviting a new hoard of parapsychological researchers.

There was another type of person, however, for whom the saints had never left the earth, those who had opened themselves to the love of the saints and who seemed to live in their essence, surrounded by a very real sense of grace. It was often said by such persons that to think purely on a higher being was to be in that being's presence. The degree of contact, however, was not necessarily dependent on one's own purity or power. Often, it was a question of mercy, or a favor from "an old friend." For every soul, it was said, had specific guardians who watched over its development, and an attraction to a saint or master from a seemingly foreign culture might actually be revitalizing a channel which had been with that soul from long before.

W.B. Yeats had written that, since the Renaissance, the writings of European saints, however familiar their metaphors and structures of thought, had ceased to hold our attention. Had the highbrow literary culture grown that far away from popular fantasy? Or was popular fantasy now totally wrapped up in black lace video madonnas? I wanted to detach myself from all this. I wanted to meet the saints, not through the orthodox canons or intellectual efforts to contain the miraculous, but rather, to glimpse that purity which I had seen manifest through those who in some way still lived with them.

Saint Collette

The fresh cut fields of the old Flemish North could not be much different than they had been five hundred years before, during the time of Saint Collette. The hay was stacked up in square piles at evenly measured areas. Bird calls filled the air as the fog slowly lifted from the horizon. The fields, still moist, were permeated with the scent of dung and cut hay as the sun began to lift that special smell from the earth. The air was crisp and cold. The village was clean and simple and echoed with the sounds of church bells and barking dogs. People moved like tiny miniatures in the morning. Here, the charm of the old remained.

"I sit in your chapel, Saint Collette, which marks the place of your appearance in this world. While raised in this simple abode, you came into an age of turbulence and schism. The great church of Saint Pierre, a powerful seat of religion in its day, still stands in the center of the village. Its bishops and prelates, in your time, were loyal to the Pope of Avignon.

"You entered a number of religious orders at a very young age, but none of them could contain the thirst for truth in you. Your vision sought a greater cause, one which would unify the rent Christian world. You soon retired into seclusion and remained in a small cell suspended between two buttresses of Notre-Dame de Corbie for years, not caring for any security offered by the world. There you prayed, performed penance, and sought the guidance of the Holy Spirit, and it was there that Saint Francis of Assisi appeared to you in a vision and bestowed upon you your life's mission, to reform the holy order of the Poor Claires.

"You worked with such diligence and zeal that the dubious church authorities eventually gave you their support, and you traveled throughout the land establishing new orders and reforming existing ones. Your fame spread. Many heard of your spiritual gifts and ecstatic visions and came seeking healing and personal guidance. Among them was the fiery Dominican preacher, Vincent Ferrere, an inspiration to thousands, yet racked with anguish over the schism. He found comfort in your association and left his holy cross in your keeping after your prophecy as to the little time left in his earthly sojourn. It is said that on the day of your death, the schism ended....I sit in the aura of your holiness, Saint Collette. I sit in prayer and in longing for that unity of all souls in Christ, that unity which you made your own."

In that small church, in the small town of Corbei, one could see as clear as the air the invitation to die. And yet, it was so far, so pure, even while it beckoned gently; its resolution was as powerful and immovable as the silent cathedral stone.

Joan of Arc

A modern tapered crucifix now rises on the exact spot where Joan of Arc was burned. Towards one side of the square is a church of ultra modern art with long sloping sides, sunken floors, and visible wooden beams. A covered market place has grown up around the monument. Everything is clean and new by the smoothly paved area. Vendors sell grapes, nuts, olives, and other summer produce.

Saint Joan, in her childhood, heard the voices of Saints

Michael, Margaret, and Catherine. She believed in her
voices and she followed them. As wonders and miracles
occurred, others also followed. Some say that the voices
stopped because her mission had been accomplished. The
Burgundians and the English, however, thought differently. According to the Apostles' Creed, she was a heretic,
for the direct command of God could only be received
through the Church. As she burned, she held her crucifix
and looked towards heaven.

The market place is so captivating, so multi-dimen-
sional, that one forgets the peril of the path is real. In
days of relics and infidels, one who heard an inner voice
was a witch or a heretic. Now, one is psychotic, or at
best, subject to "ego-inflation."

I once had a friend who from an early age was convinced that he had been one of the twelve apostles. As
years passed, he grew more sure of his conviction, for it
was constantly being confirmed by signs and visions.
Then one day, he made a new friend who from an early
age was convinced that he too was this same apostle.
What happened? Was the weaker one relegated to a new
role? Did belief systems collapse?

And even if one's inner myth finds acceptance in the
world, what becomes of the inner life? How does the
psyche mask its own doubt? It is said that towards the
end, Saint Joan wavered. Had she been deceived by her
own voices? Were the guidelines of the Church truly
drawn up in good faith to provide boundaries of sanity
for the social order? Were the statues erected to the saints
an intrusion of pagan idolatry, or might they be monuments to the stinging conscience of the world which has
always looked with deep fear on those who live their vision without compromise? Perhaps such identifications are
not entirely literal. Rather, they are a way in which the
soul may revive an aspect of itself, an aspect which is
meant to live in a glory which is sadly absent from the
visions of this world. The present consensus of sanity now
stands by the doorway of thermonuclear war. The crucifix tapers upwards into the unknown.

43

Lisieux

The rain fell lightly over the grey skies of Lisieux, the sanctuary of Saint Theresa, God's little flower. The air held its own quality of determined urgency. Up on a long hill was a complex of churches, monuments, and museums. Down in the town, the places which Theresa had frequented were marked off as sites of pilgrimage. Stores were selling all kinds of pictures, amulets, statues, and photograph albums. Every twenty minutes or so a new load of tourists and pilgrims came in on another train.

Ignoring the fanfare, I entered the chapel and tried to open my heart to Saint Theresa. I told her how confused I was about the inner life and how difficult it was to persevere. You prayed and nothing happened, but you tried to convince yourself otherwise. This was known as belief. Perhaps the churches, altars, and statues were another escape, another form of going to the movies. I sank into a silence and into a gloom.

On leaving the chapel I felt myself drawn to a young priest who was standing by the door. Hearing that I was a traveler from the United States, he invited me in to sit and talk for awhile. I felt a great cleansing in the presence of this young man. We spoke of various experiences on the "path" and of the feelings of despair which came at times. "We cannot climb the mountain by ourselves," he said softly, "but Christ can heal us through prayer and surrender." I had heard it all before but this was different. There was no one trying to push some kind of love or religion over on me. He was just there.

He spoke about prayer, how Saint Theresa had called prayer an *"élan du coeur,"* a literal bursting beyond oneself, a throwing of oneself towards God. "Real prayer," he said, "is when we enter into a heart-to-heart relationship with Jesus in complete trust and confidence. I do believe that in this way there is no problem that cannot be solved, that all can be done in His name." We said goodbye, and he wished me luck. I continued to wander through the shrines.

France

Saint Theresa had that complete confidence. "I shall spend my heaven doing good on earth," she said. "After death I shall let fall a shower of roses." And her presence was there, like a soft, soft flow. I had often wondered if the saints still spoke; not to the one visionary, but to the flocks loaded with the burdens of the world who lit candles in church corners and came seeking favors. Did they look down when the offering was made, or appear in dreams to give needed direction? Perhaps one looks too far. I felt that on this day Saint Theresa had let one of her roses fall on me, through the words of a young priest. But how difficult it was to accept the gift from another.

The sun was now coming through the clouds. I took a train back to Rouen and watched as the fan-like rays beamed out on the countryside. Looking down, I noticed that all my clothes were filthy. I took everything off except a blue bathing suit, walked into a laundromat in town, and threw my clothes into a washing machine. The detergent dispenser was broken, so I asked a woman, in French, if I could borrow some detergent. "Why, of course," she answered in perfect English.

Her name was Nancy, like the city in northern France, except that she was from Illinois. She had been in France for some years working towards a doctorate in medieval history. "When you get your clothes back on," she said, "you can come up to my apartment and have a cup of tea, if you'd like."

Her small studio felt like a cloister. Nancy had put on a record of medieval lute music which floated through the air as I looked out the window at the cathedral spire. We sat and drank tea by a glass-topped table. "You know," she said, "if you were a European I couldn't invite you up here. I wouldn't even smile at you in the street."

"That's all right," I said, "I try not to expect anything."

"Well, it's good to see an American," she said. She

wanted to know what was happening in the States, and I told her, "nothing." She said that she thought as much, and that since her junior year abroad as a college student she had no further desire to return to the States with its fast food. I wondered if she was aware that MacDonalds was now on the Champs Elysées.

Suddenly she became very serious and, turning towards me, asked me how old I was. I thought about it as the music floated through my ears and my eyes fastened on the spire through the window. "At least five hundred years old," I answered.

She didn't take me seriously and went on as if I'd said nothing. "I'm thirty years old," she began to complain. "Do you know how that feels? You can't believe it until it happens to you. But I really don't care," she covered herself. "If I die now I've got no regrets."

Things got quiet again. I looked at her closely, seeing both a young woman and a very old woman through her eyes and hair. Our breathing was going in rhythm, mixing with the lute. There was nothing to say or do. Still, I felt connected with Nancy, like we were just supposed to be there. I didn't think I had to play Henry Miller and try and make everybody, or Billy Graham and try and save everybody. I didn't have to play any role at all, and there were no explanations necessary. I felt fine drinking tea and being at least five hundred years old.

We exchanged addresses and said goodbye. As I walked out the door I heard the word "completion." There was an all-night train going west. I could be by myself and watch the world roll by out the windows. I went down to the station and took the train to Vannes.

The Cross at Vannes

There are wars and nations, and old schisms returning in their new garbs as ideologies, but in the aura of the saints neither life nor death matters. Their presence itself speaks of a beyond, and the gift is imparted through taking refuge. In refuge comes transmission, a silent gift which can never leave you.

The cathedral of Vannes was supposed to contain the bones of Vincent Ferrere. I had planned to walk over to the cathedral and pay homage, but instead found myself walking in another direction, clear out of town.

Jesus lay, stretched out on the cross, superimposed on the city, on the vendors by the port, along the green thicketed canal. A voice spoke, "The death of Jesus is thine own." The presence increased. From here, one could go no further. This cross was the juncture, and all endeavors were but useless efforts to postpone its inevitability.

The canal headed towards the open sea, but a small path cut through high grass and thorned bushes leading to a plateau of an open field. One had to die. The great lie of living was unbearable, but the desire for death, itself, shut the door. The level field, where I wound up, had not been visible from the canal side. But now, in the distance, I saw an enormous crucifix. It was about fifteen feet tall, made of solid stone, with a circle around the center. Carvings of the bearers of the gospels, the lion, bull, eagle, and man, were symmetrically placed on each side.

Here was the full presence, and here was the understanding that through his death alone would death be possible. I slowly approached the cross and read the inscription on the stone:

> Those who are dead upon the cross
> are eternally living in Christ.

I felt a dazed relief and sank down by the crucifix, opening to that sense, that shock, that one is not the author of one's life. Vibrations entered from the master realm:

The Cross at Vannes

To surrender thy life unto the cross
and to die into its glory has been the
challenge of mankind since the beginning.
For, from within, it is known that all
roads will ultimately lead to the cross.
It looms over the world as a shadow, and
beckons to mankind through the ages.
In thine own shadows would thee forget
and attempt to enjoy a world which is
but a shadow of that truth, to take of
those things which are by birthright,
already thine. In truth, they are all
meant to lead thee to the source, to
the wellspring of love.

I sat by the shrine, feeling my brow soothed and my
soul calmed, and I prayed to remain conscious of the
cross and of the life which it holds that lies further than
any eye can see.

Lourdes

The night train drifted down the coast towards Spain and
arrived at a Spanish port near Hendaye at dawn. The sky
was filled with the spirit called "the World," and the
ocean air held a vibrant intensity. The sea itself was
clammy and cold. Gusts of wind blew through the thorns
and brambles which grew out of the rocks along the har-
bor. New vibrations were all around, shooting out of the
rocks, the earth, and the air. There were strings of white
stucco houses, homes of adobe, and the portside market
was already active, filled with oranges and olives. I spent
the morning wandering through the smell of the market
and the seaside air and thought of that old rambler,
Whitman.

France

At a certain moment, maybe it was the brine and sway
of the sea, his soul came close, and I wondered aloud,
knowing that he could hear me: "The wanderer, who is
he? Watch them pass; blond-haired Norsemen, dark-eyed
Spanish women, cathedrals so dark and light together.
The rolling elementals, how healing they are! Were you
(or anyone else) ever really free in them? . . . . Still, you
gave yourself, and I am sure that you walked empty and
hungry like I do, and that you peered at the passing
faces, looking. . . . The herdsmen of Catalonia, I saw them
in the distance, their flocks grazing on the mountain
which rises into the still point between the mind's motions. And you knew the healing of the sea and air; and
the earth. . . were you taken up with her fruits and running juices? I am sure you were all of this; in fulfillment
and wanting, in peace and in suffering, you always sang
the music of your life."

The sun stretched over the rocks. Wanderers seabathed,
leaving their knapsacks in the sand. I remained for the
rest of the day, sitting on the rocks by the water until the
sea winds came strong, reminders of the train and the
journey.

From a distance the town was pure commerce. Row
upon row of cafes and hotels were interspersed with souvenir shops selling plastic bottles in Mary's form for collecting holy water. Every conceivable kind of religious
trinket was on sale: wooden carvings of Mary and the
grotto, thirty different kinds of rosaries, Christs, Madonnas, silver covered Bibles, story books, pendants, and
more. I was already tired from traveling, and the whole
thing got me disgusted. "I'll just hurry down to the grotto," I thought, "get some water, and clear out."

I quickly made my way down the hill of curving
streets, passing hotel after hotel, and finally arrived at the
main road which led to the grotto. The sun had gone
down into the mountains.

Lourdes

As I passed through the iron gates, I suddenly saw throngs of people massed up ahead. The entire area was filled with humanity. It was unbelievable. There were lights everywhere. Moving closer, I saw that almost everyone was carrying lit torches. Still feeling cynical about the whole thing, I pushed through the crowd, remembering that these torches had also been on sale in all the stores.

While trying to make my way through all these people, however, I became aware of their gentleness and sense of vigil. This was not just a massed Broadway crowd. This was a procession. I raised my head to try and see the grotto where the Blessed Mother had appeared to Saint Bernadette and for a mile straight saw nothing but realms of people holding up torches. My pulse began to quicken as my heart beat faster. This enormous crowd of pilgrims, I had never seen anything like it, not even in India. It was as if all humanity had come out this night to pray to the Blessed Mother. I felt humbled, not humbled before God, or before some mystical insight, but before humanity itself expressing devotion to some intangible event, a miracle that was said to have taken place long ago, that still captured the heart of the world.

Even now, some threw away their crutches and braces as they partook of the holy water. This was humanity coming en masse towards this holy site, towards this awesome event, and I had put myself above it, wanting to get ahead, get some souvenir water, and get back to my train. All that I needed was already surrounding me. I silently thanked Mary for this happening.

The procession moved slowly through the evening and on into the night, circling around the grotto and singing "Ave Maria" and "The Bells of the Angels." With each chorus the language would change, and the songs were heard in French, German, Swedish, Italian, and more. The mass of humankind began to congregate in the center as representatives from different nations were called. An unspeakable joy spread throughout the area where all had

51

come together to demonstrate their faith and love. The Blessed Mother was addressed as the "Queen of Peace," and prayers were offered for peace on earth. Those on crutches and in wheelchairs came up for prayers and blessings. As the prayers rose in the night air, the statues of the saints and angels seemed to come alive, as if they were the guardians of the race and of the faithful.

This gathering would not be carried by international television to homes across the world, but the message would still go out. The pains taken to make the journey here, the vigils, songs, prayers, and torch lights; all this would be taken up by the angels and sent to those in need. The energy created this night from the outpouring of thousands would remain with the earth to guide her children.

In language after language they sang, and the singing came from the heart. I kept thanking the Blessed Mother for this night, for opening my heart, and for showing me my blind arrogance. I arrived at the grotto, took some holy water in my hands, and sprinkled it all over. I held on to the water and prayed. For the first time in years I felt a faith and a possibility for this world and its people.

I laid out a sleeping bag across the river in view of the grotto. Voices lingered on, carrying across the rushing water. A train passed, leaving its trail in the river's sound. The air was chilled and misty, but I did not mind. I was sleeping by Mary's shrine, where she had revealed her form to a small child, and where she cared enough to appear and reappear.

> Voices risen in the nightlit heaven
> sounds of peace and purity
> as the universe rejoices
> in the procession.
> The faithful marching throng,
> the ever human longing
> a moment of communion held
> precious in eternity.

The Miraculous Tomb

Well past midnight a car dropped me off on the highway some miles outside of Perpignan. I walked along the road with my thumb out, having no idea where to spend the night. A car came rushing by, rustling the tall grasses along the edge of the highway. Its headlights lit up a sign, "Camping Three Kilometers." I walked quickly through the night. The air was filled with humming sounds of crickets and other nocturnal creatures. I kept walking and soon came to the campsite, hopped over the wall, and found a place to lie down.

I awoke to a dream of a long arched cloister. It formed a square like the inside of a courtyard. That was all I could remember. I took a shower and was out on the road with the sun. Just before leaving the States I had spoken with Hilda Charlton about my upcoming pilgrimage. She had playfully shown me a small article from the back pages of *The National Enquirer* which someone had recently sent to her. The clipping spoke of a small village in France named Arles-sur-Tech where miraculous healing waters had been flowing for years from the tomb of two martyred saints, Abdon and Sennen, from the early Christian era. When the first driver who picked me up mentioned the Tech River, I asked him if he had ever heard of the town. I was there by sundown.

Arles-sur-Tech is found among the curving mountains of Catalonia, separating France and Spain. Rivers of healing mineral waters run through these mountains, and many thermal bathing stations are found along the river's way. Arles is so small, however, that it is not found on most maps or tour guides. Walking into the town, I asked an elderly man sitting on a bench with a cane and thick grey hair if he knew anything about the tomb and its healing waters. He gave me a weird, "Go away, sonny" look and told me that the place no longer existed, that it had been closed for years. Something told me not to take no for an answer. I walked into the center of the town, asking one person after another until finally a bunch of kids led me into a courtyard of a large church which looked as though it had been a monastery at one time. I walked through the doorway and did a double-take. This place was the cloister from my dream.

A few minutes later, a door opened and a plump, white-haired woman dressed in the style of the old country appeared. The moment I saw her, all I could think of was "Mom." She spoke neither French nor Spanish, but only the local Catalonian dialect. We managed to communicate anyhow, and I was soon sitting upstairs at the kitchen table being fed cookies and tea. All this woman wanted to do was feed me and feed me. She kept patting me on the head and signalling that I should wait. Later, her daughter who spoke both French and Spanish arrived. We talked and I was invited to stay.

Apparently, miracle upon healing miracle had taken place through the grace of this mysterious water. The family showed me piles of written testimonies and took me out on the balcony before dark, pointing out where I could catch a glimpse of the tomb. Saints Abdon and Sennen were killed by the Romans in the first century when they refused to abandon their faith and worship the gods of the Roman emperor. Formerly noblemen themselves, they had received the Holy Spirit and were arrested as they attempted to gather the remains of tortured Christians in order to give them a proper burial. The emperor had a hard time putting the two saints to death as the lions wouldn't comply, and they finally had to be killed by a hoard of gladiators. Somehow or other, the remains of their bodies found their way to France and to this abbey at Arles-sur-Tech.

Throughout the next day, Mrs. Fernandez, the white-haired woman, kept me seated either at the table or on the balcony, as she brought out one dish after another; French fries, squash, milk. She couldn't give me enough. The night before she had spied holes in my jeans, which weren't too hard to find as they were situated around the seat of my pants, and had insisted that I give them to her to sew. When she wasn't cooking she was out on the balcony sewing and smiling. I couldn't look at this woman without experiencing the "Ma" feeling, which was fine as long as I knew I could leave at any time. It was also nice to have cooked meals.

It turned out that ever since that article had appeared in *The Enquirer*, these people had been inundated with letters and phone calls from America asking for water. The problem was that no one in the house understood English. So I went to work answering letters and taking calls from as far away as New Haven. God knows how they found the phone number.

Mr. Fernandez was the caretaker of the church. He was brittle and shrunken frail and walked slowly with a slight limp. Later in the day, he led me through the cloister and on into a grey walled sanctuary which was filled with large, hollow statues of saints, various altars, and black iron candelabra crudded with wax from burnt candles. There were a few wooden benches by the altars, all lined with a thin film of dust. While the Americans knew about Arles-sur-Tech, the French had apparently not discovered it yet. Outside the mausoleum-like sanctuary was a small courtyard surrounded by an imposing iron gate. Walled in behind the gate was the tomb. Mr. Fernandez tried to explain the ritual of the tomb and when one could collect the water, but I didn't have quite the same rapport with him as I did with his wife and couldn't really understand how the whole thing worked. He went to the gate by himself, picked up a bottle of holy water from a row, and gave it to me.

As I was about to leave, Mrs. Fernandez came out with a bag filled with fruits, bread, and cheese. It was somewhat easier for me to understand her speech. She said something to the effect that every night before going to sleep she felt that she had not quite done enough for God during the day. Then she suddenly burst into tears and asked me to remember and to pray for her.

France

I left France that night by train. The first leg of the pilgrimage was completed, and still the medieval dream hovered somewhere, a floating shadow, a longing of the heart for its ancient, unmixed conscience.

2
Italy and Greece

Genoa and Florence

The train droned on, lulling the senses to its rhythm.
Sleep was broken by the clamor at each station. Environments would now be a greater challenge. I did not speak
the language, and there were no more old friends to stay
with.

We arrived in Genoa at one thirty in the morning.
Everything was closed down. I didn't care but kept walking. Then I saw a hand beckon from a rising stairway.
The man only spoke Italian, but I understood that he was
offering me a place to stay. I followed him up the stairwells and down the streets in the dark wondering what
he wanted and hoping it was only money.

Half an hour later I found myself in a small, comfortable hotel. I could say that I followed my hunch, or that
in fact I had no choice (if there was any difference between the two), but it turned out to be a straight deal.
He gave me a room for a modest price. We tried to
speak, but it didn't work. But who cared? For now, it
felt right. If the vibe changed, I'd get right out. A wanderer had to take such chances.

I looked out the window before closing the shutters.
The streets were still. Everything sat in silence. I mental-

ly traveled back through the day to smooth out any kar-
mic wrinkles. What was the world and who knew? One
thing, however, was clear. Without that openness, that
willingness to risk direct contact, everything would come
as secondhand information, always.

In the morning street noises came through the window.
Barking dogs were being walked on the cobblestones outside. The clanging of breakfast forks, auto sounds from
the highway, and sounds of living men and women rose
through the city. Sound superimposed on emptiness, the
fear of the Lord, was the beginning, the awe at the vastness, at the mystery of each morning.

I sat alone in the little room and opened the shutters. I
had no one, and that's when you really have to look in.
In meditation the Holy Spirit was like the light of dawn
and more, the mystery and the source of light. Like the
sun, baptism could renew itself each day, could descend
each day through the flow of events to the ineffable. In
the space of pure aloneness came the renewal into the All
which knows nothing. The present Lords of the Earth
with their money and machines were but sorely separate
reflections of the All. I felt as much pity for them as for
the lost and lonely, as for my own gaze in the mirror.

Florence

An aura of inspiration surrounded Florence, city of
beautiful works. That inspiration filled the streets, the
rivers, and the monuments as the pulse of a place where
expression had erupted, where beauty and life had been
carved and painted into celebration.

For hours I stared at Michelangelo's Christ being taken
down from the cross; the deep power which seemed to
surface only with suffering. Museum upon museum,
gallery upon gallery, layer upon layer, age upon age.
These monuments of earth, stone, color, and language,

Patriarch at the
Cathedral of Florence

could they ever rise from their own ashes and point
beyond themselves?

The history of culture passed in procession through the
gallery walls: Christs and Madonnas, classical columns,
sculpted chests and brows, all larger than life, all
definitive vision. In the modern galleries the visible world
was blasted open, the overshadowing ideal gone with the
image, like a great rug pulled out from under the world
and feet falling in a maddened dance for balance.

I had to admit, I felt more comfortable in those rooms
with the smooth marble pillars, with their heroes and rising columns. I was always hoping that somewhere, on
some pilgrimage, I would look out the window and see
the flag of the Golden Age raised for all nations to see,
but such is the hope of the dying. Those cast off the
bedrock of cultures and races, ideologies and gods, would
have to swim out to sea and hope to arrive at the unique,
at the all-embracing, or maybe not arrive at all. The
heart knew that faith which was greater than the hoisting
of any form. It was filled with firmness as well as with
fear, with confusion and with generosity, with the enor-
mity of suffering which humans hold, and with the compassion to contain that suffering without bitterness, to
walk freely upon the void.

I walked out into the streets, museums in the making.
They were filled with vendors, bake shops, panhandlers,
artists, souvenir salesmen, and tourists. Crowds swelled
out over the bridge between the two sides of Florence.
What quality could stand up to those crowds, the crowds
of streets and ages? Surely, it could not be a turning
away from them. There had to be something else,
another quality, one willing to claim uniqueness and willing to abandon to nothingness.

Attention danced around the world, gnawing after the
endless flow of changing objects. How could one pull
away from them? Was there another chastity, one wed to
the absolute being, to the truth of aloneness, which could
yet ever open to the ten thousand things?

Le Louppiano

Outside of Florence, situated on a mountain in Le Louppiano, is the international home of the Focolore Movement founded by Chiara Lubich. Some Argentines who were on their way up the mountain gave me a ride. In twenty years, Le Louppiano had grown into a small city of over eight hundred persons dedicated to the principles of love and unity through Christ. The whole area exuded unity. You could feel it in the air. I asked if I could stay for a few days. I was given a room and a smile. There were no feelings of suspicion. No one asked what my beliefs or doctrines were.

Only a few people in residence spoke English, but after a meal, two of them took me on a tour of the place. There were numerous buildings on the near side of the dining hall with all kinds of projects going on, while the rolling hills on the back side of the mountain were filled with crops. Emerson had refused to join Brooke Farm, and Thoreau remained in his Walden retreat. Even when in jail for civil disobedience he was alone. As a law of nature, however, the solitary walker would fall. What could perhaps remain was the transmission of inspiration, and that inspiration belongs to the community.

Now I had lived in a number of communities, and I knew the pitfalls. It was easy to get lost or ingrown. You might think that you had kicked the policeman out of your life only to find three new ones standing guard by the back door. Would it ever be possible for individuals to come together for some other purpose than the necessity for survival or the mutual assurance against exploitation called justice? Could creative aspiration manifest a higher bond? Centers of visionary energy were rising on the planet. Would they become free enough to flourish, or would they be trammeled down, or smother themselves as so many others had done in the past?

The Focolore Movement rose from the dust and ashes of the Second World War. Italy was devastated. Her air

raid shelters were not as sophisticated as in other countries, and imminent death was a possible reality for all.
In these shelters as the bombs fell, Chiara and her friends
began their work.

Chiara has explained that prior to this time, although
leading pious lives, she and her friends had their own
particular plans and goals to fulfill as the foremost thing
on their minds. One wanted to become an artist, one a
teacher, and another a full-time student of philosophy.
During the war-time desolation, however, they found
themselves bereft of everything and confined to the radius
of an air raid shelter. With all plans crumbled into the
likelihood of there being no tomorrow, the present moment became the only reality.

A woman would be in the corner of the shelter huddl-
ing her five children around her. Chiara and her friends
each took one of the children, lessening the burden. Soon
they began to distribute food in the shelters and perform
other needed services. Here the essential was discovered.
When all else had fallen away, there was only love. From
that time on, the girls decided to live for love by putting
the words of the Gospel into practice, "Love one-another
as I have loved you." What effect might there be if only
a few actually began to live these words? Here, the effect
was the beginning of an international community of souls
living in a mutual pact of love and charity.

The word "focolore" means "gathered around the
hearth." The members of the movement live either in
small groups called "focolores" or in their own families
while seeking to manifest the unity of God's love at every
moment. This mountain community spoke more forcefully
than all the preachers and their slogans. Here was love
made visible.

At the dining room that evening, people were anxious
to speak to me as a new guest. We talked into the night.
As in many communities and ashrams, there were the
hard-core faithful, the curious visitors and relatives of residents, and the usual hangers-on, the "fringies" who
maintained a peripheral relationship with the community.

Le Louppiano

This last group was often the most interesting. One young man who spoke English explained to me that he dropped in from time to time to "get his head straight." "The food is good, and they don't work you too hard," he said. He warned me not to trust anyone since certain areas of the country were full of trained thieves who were so skilled that they could cut a hole in your sleeping bag and steal your money belt while you slept.

I spent the next few days in the woodworking shop where artifacts were designed and manufactured by the residents to be sold in stores. After being shown around, I took up a station by the sanding machine. I soon learned how to carve wooden toy animals from a pattern. The workshop was alive with vitality and cooperation. It was incredible that after a few hours I wasn't dead bored as I had been in all my office jobs. Everyone was eager to share, and I soon became familiar with the entire manufacturing process. There was no punch clock, no foreman policing the workers, no labor disputes or heavy sense of drudgery. The finished products exhibited true artistry, with care and attention to detail given to each piece. Of course, the work demanded certain sacrifices. The idea wasn't to do just what you wanted to do when you wanted to do it, and there certainly was not much of a profit motive.

I spent evenings with a French-speaking resident named Antoine and an English-speaking Malaysian named Phillipe. They took me on tours of various work projects in the community. In one shop electric meters were being assembled and sold to the city. In another camper-trailers were being built under contract from a private firm. On the far side of the village, there were studios where clothing was designed and made by hand. There were working areas for sculptors and artists whose works were on display in a gallery. Phillipe took me past rows of paintings, woodcuts, prints, and handmade post cards to a piece of abstract sculpture in which a large smooth stone was turned inwards with a smaller stone curved over it. He explained that the piece expressed unity,

63

showing how one must be willing to bend towards the will of another as in the bending of the stone.

When the sun was down and the air cooled, we would walk through the vineyards. With my interest in communities, I fired question after question at my friends. They didn't seem to be annoyed at all and were willing to share the communal structure with me. Their answers flattened ideas bred into me that the spiritual could not be practical as well. Everything at Le Louppiano was well managed and organized to the last degree, and yet one did not feel that individuals were being sacrificed.

Most of the young people were living out two-year periods of residency as students. They lived together in small focolores and shared everything from meals to meditation. The families in permanent residence had their own dwellings and were more involved in managing the land itself. Different committees took on various responsibilities, from finances and work projects to construction and community relations. Once a day, at noon, everyone would assemble for Mass, renewing the sacrament of fellowship and love.

I slept in one of the trailers at the edge of the hill, never bothering to lock the door. Late one night, I became aware of a presence. Then I felt a hand slowly easing towards my waist. I had been traveling alone for quite awhile now and would get lonely from time to time. Maybe this was my chance. Maybe a woman was sneaking into my bed, although I hadn't noticed anyone giving me the eye in the dining room or anything. But one had to be careful. Some years before I had been sleeping alone on a cot in a temple at Surya Kunda, a small village in Northern India. In the middle of the night, a feminine figure had softly entered the room and had slowly run her hands over my body. I instinctively reached out to draw the figure towards me when I looked up and found myself staring into the face of the ugliest man I'd ever seen. He was the village transvestite who danced and participated in some of the theatre groups in the area. I gave a startled yell. He jumped and ran out of the room but returned and tried again two nights later not wanting to take no for an answer.

Then the memory of the warning about the thieves appeared in my sleepy consciousness. Luckily, I was a light sleeper. I felt the hand searching for my belt which I, like so many other travelers, wore around my waist and which contained my papers and whatever money I had. Most of the time I just kept the belt and all of my other things in a small shoulder bag. But it was wise for the traveler to be as careful as possible so as not to encourage anyone else's bad karma. I leapt up yelling in French, "*Voleur, voleur,*" (Thief, thief,) and lunged at the shadowy figure, who bolted out the door at amazing speed. The lights in the neighboring trailers all went on. Soon there was a crowd around me asking all sorts of questions. I told them to forget it and went back to sleep, this time locking the door.

The community residents were shocked when I told them what had happened, but my English speaking "fringie" friend gave me an "I told you so," and pointed out that the trailers were right on the border of the land, and that anyone could have come up during the night.

Still, the optimism and joy which pervaded this mountain remained unruffled. It reminded me of our beginnings at the farm. Once two truckloads of French gendarmes came and carted us all off to jail in the middle of the afternoon. "It was just an investigation," they said, but we loved it. At that time, being on an ashram was like being on a permanent vacation. It wasn't until the novelty wore off that individual desires and destinies began to assert themselves, creating tension in the community. Sooner or later, every intentional community had this problem. Those that were able to be flexible often survived but ran the danger of the assimilation and loss of their original ideals. Others, such as the Cathars of 13th-century France, burned together while singing their hymns. These small focolores were close-knit, and the people seemed to exhibit a genuine concern for one another. But what would happen, I wondered, when some of the members began to hear their own music?

Phillipe and I walked through the vineyards on my last evening. They were vast and full of life. All sorts of insects flew around us, and their night sounds rose from the fields. The grapes, a powdery blue, were already hanging heavy and ripe. The rows seemed to extend outward as far as we could walk, deep blue on green, abundance everywhere.

Phillipe spoke of his own experience of unity. His soft voice had a certain benign quality which I had come to recognize as belonging to surrendered souls. Although raised in Malaysia, his family had given him a Western education. "I try to listen to everyone," he said, "as if that person was God himself. It was not always like this for me. I did all kinds of things when I was younger, trying to find something through your American way of life. In my last year of high school, while studying the Western classics, I came across the Greek word *ekstasis*. My teacher translated it as 'moving outside of oneself,' and it has always stayed with me. This is what I am trying to do here. I do not have to go into solitude to find God. I surrender to Him through everyone I meet. That was the beginning of my inspiration, and I feel it very much here, to live by saying yes to God through every situation."

To say yes, not only alone but with others, one may be cynical until one has seen the people of the way as they rise in the way. I was surprised, but not too surprised, that on the morning of my departure, when I had gotten up early to get a quick start, Phillipe and Antoine were up and waiting to drive me into town to the station.

Assisi

The train arrived in the evening. I walked to the village from the station, buying Italian bread and some cheese in

Assisi

a small grocery. I sat in the park and watched small bands of pilgrims pass through on their way home. The good Sisters gave me a bed in one of the empty dormitories of their hospice building. Everything was closing down for the night. I walked out to the main road and hitched three miles up the hill to the city which was the central site of pilgrimage. There, sitting on the wall outside the cathedral, I looked out at the night-lit homes forming patterns of light against the darkened sky and mountains. I took out the rosary beads given to me by the priest in Marseille.

As things became quiet, I looked around and felt myself to be in a holy city, a city of God in another age. The wall line led upwards to the rising steeple and tower. Saint Francis, his ministry arisen from an irresistible call, threw away his clothing of designation, his father's wealth, and with his pure joy chose to live by the Word alone. There was strength. And there too was the most pressing question, the question asked while kneeling by the darkened crucifix at San Damian, "Father, what shall I be?" There was silence between the beads of the rosary, and the waiting for an answer which could not be invented.

I had hardly noticed a young man who appeared by the wall. When I looked up and saw him, he silently came over, walking slowly to the ledge. His name was Remo. He was born in Rome and now lived in Assisi. He asked where I came from and if I believed in God. His shirt hung out of his pants. There was a resignation in his voice as we talked, walking by the side of the wall. He said that he had lost his faith. He had left his family, and now lived alone in a room not far from the center of town. He cleaned bathrooms and tourist sites for a living. It was getting late. He invited me to his place, but my things were all down at the hospice, and I had to catch a ride before the cars stopped altogether. We agreed to meet for lunch at his apartment the next afternoon.

Early next morning I went to Mass at the original chapel that Francis of Assisi had built himself. The atmosphere was starkly vibrant. The aura of purity and zeal was all-pervading. People still worshipped the saints and collected relics for protection. But there was a certain gap in it all, a nostalgia for what these places might have once meant to the world. People worshipped saints, but inside, no one wanted to be like them. The sites of pilgrimage were a little too lofty, a little too lonely, and hope in the invisible diminished as the accelerating world multiplied objects and apparent possibilities. The monastic ideal was gone, the blurred remnant of a dream in an age which had passed it by.

While traveling through France, we had once stayed with a group of Trappist monks for awhile. They weren't allowed to speak with one another, but were not prohibited from speaking with visitors. They talked to us constantly. One evening we made some sweets and gave a few to the monks who attended the hospice. For the next two weeks they came every night for more. But they lived by the old order of things. Every night when we arrived at the monastery, we would find a special meal prepared for us. And one evening two monks came to our room and asked about our theology. In the end we prayed together. I used to watch one who, every day, walked nervously by the wall, probably dreaming about what he had left behind. And there was the old one with the long beard. He had that gleam in his eyes, and you knew he had broken through. He was the gatekeeper and would greet us every night with a long drawn out OOOMMM.

Before leaving, I was granted an interview with the abbot of the order. We met in one of the stone rooms near the chapel, which contained a long wooden table. We both hedged around the table as we hedged around with each other, not quite sure who would play what role. When he finally said to me, *"Nous sommes tous les petits riens qui cherche la lumière"* (We are all little nobodies searching for the light), I lowered my head and didn't answer. We were both lost in our play-acting religions. I just thanked him and said good-bye. I never forgot that man and that moment. It brought up the memories of

Assisi

God knows how many lives in God knows how many monasteries. And to what did it all avail, to come back again and try in another form? It was like being traded to another baseball team.

At noon I came to the church of Saint Claire. I was at the altar when the bells began to ring signaling the closing of the doors. I remained there in meditation. I heard the doors close, but I didn't care. I could remain there all night and just feel the inner comfort of not having to try to arrange reality in any particular way. Sometime later, I got up and walked over to the heavy metal doors. One of them had been left ajar. By whom? It didn't matter.

The streets were empty now. I walked through the sun-bleached town and rang Remo's bell. He opened the door and took me up a series of stairs and landings to an attic apartment. He had one small room, unpainted, with a window from which you could look out and see down the mountain. We sat for awhile, and he said it was good that I had come. He asked if I liked pasta. I was never one to miss a chance for food, and nodded, "Yes." He washed out some faded plastic cups and poured some warm wine. There was a time when, if I had been invited anywhere by anyone, I would be anxious and hopeful that something eventful would happen. Maybe the meeting would be very mystical and we'd remember one another from a past life, or very religious and somebody would be saved or healed, or very erotic and we'd make passionate love. Now I could accept that, aside from going through the rituals which reinforce our separateness, nothing ever happens at all and nothing ever would.

There was a makeshift table where he set out the plates of pasta. I sat on the edge of the sofa and he sat on a box that stood on the bare floor. Empty bottles placed on the ledge one after another lined the walls. On the other shelf there were a few more blue plastic cups, two cans of string beans, and a rumpled plastic bag of pasta. As we ate, two flies buzzed around the table. My friend's hollow eyes and thin nose spoke of tiredness, a tiredness which he totally accepted.

Our conversation floated through the air like the dry particles of dust. In the empty room, even without questions, all was visible. The weak fall from this world. The pretty face, the quick mind sends you into retreat, further and further, until fear hardens like crusty old clothes. Perhaps you hold on and frame yourself into some kind of life to keep from falling. Particles of dust lit by the sun floated through the room. I remembered the bars in New Orleans and traveling with drunken hobos in box cars. I remembered reading Sartre in a laundromat. At the age of nine he had given up on God. "My God became literature," he said.

We said our good-byes after lunch. I was relieved to get out of there. There was no one to save and no one to love, no one to agree or disagree with, just the dust floating through the air of the empty room. You caught glimpses of particles, glimpses of faces in railway stations, on benches, and through windows. And in that glimpse maybe you took something with you, some altered sense, some mark of recognition. You would forget of course, or more than likely, never notice at all. But there is that other part of you which would never forget, that part which has seen down deep and which, somewhere, remembers all.

The day was drawing down. I had seen just about all of the churches and monuments and was planning to leave for Rome on the evening train. But still something was missing, and I couldn't bring myself to enter the station. I knew what it was. I had not visited the church at San Damian where the crucifix had spoken to Saint Francis. They say that Francis of Assisi was praying in the dark, asking the Lord what to do with his life, when a voice came out from the crucifix telling him to "rebuild my house." At the time, Francis thought that meant the little church itself, and he worked to have it restored.

I had thought about that place all day, but it was too long a walk. Now, carrying my traveling bags, I had to start off in the heat, taking dirt roads that cut through the cornfields and the vineyards. It was at least a four

mile walk, all uphill. I picked grapes along the way for
sustenance. A couple of miles up where the road turned
by the end of a large field was a tank filled with water
which was used for irrigation. Seeing no one around, I
took off my clothes and jumped into the cooling water.
After awhile, I climbed out, dried off in the heat, and
continued towards the church.

It was a simple, stark church. I sat inside and gazed at
the cross which Saint Francis had gazed at, secretly hoping for a miracle. When I had squeezed all the emotion I
could out of that, I looked up and around at the building. There was a group of nuns praying together in silence, lined up all along one dark bench. They looked as
though they were from South America. Their cream-
brown faces were clear and serene, concentrated inwards.
The purity of their intention shone through them. The
church was dark. The walls and benches were dark. The
air was still, and their faces were shining in the darkness,
glowing in the stillness of self-dedication. They did not
look out of place kneeling by the wooden benches in their
blue robes with white borders. As I gazed at them and
then at the crucifix, I felt that I was the one who was out
of place, a buffoon in a house of purity. That feeling
passed too, like things always do.

Towards the rear of the church compound was the
room where Sister Claire had lived for some forty years
without ever going outside. The very same wooden
benches and tables were there. Hundreds of years before,
the Sisters had taken their meals and lessons in this room.
All was absolutely stark with nothing superfluous, and
through this darkened silence of self-mortification came a
peace which only the Holy Spirit could bring.

While making a last stop in the church, I had the feeling that something was missing from my body. I felt
around my neck and noticed that my wooden beads were
gone. A chill passed through me, not so much because of
the beads as because of what I thought losing them stood
for. It was definitely a negative sign, a serious warning.
My mind began to manufacture all sorts of offenses for

which I was guilty. The Lord was displeased! I had not
been paying attention to my spiritual practices. I went on
like this for awhile. Then the bathing tank flashed
through my mind. I walked all the way back through the
fields lugging my baggage. The sky was darkening. As I
dragged myself and my belongings past the vines and
grape clusters, I castigated myself for being a fool, a
thief, a false enjoyer of an illusory merry-go-round in the
name of a so-called pilgrimage.

I arrived at the tank and looked in where I had
bathed, scraping the bottom with both my hands. I
looked all around the sides of the tank, under the leaves,
and combed the entire area where I had left my clothing.
I found nothing and felt that sinking sensation in the
solar plexus. As I picked up my bags and got ready to
leave, I silently called upon Saint Anthony and also upon
Ma Parvati Ammal, the miracle worker of Ceylon who
was always very good at finding parking places. As I was
walking away, a thought flew into my mind. It literally
flew, as if someone had sent it winging down on an
airplane. "Wood floats!" Startled, I put down my bags
and scanned the entire surface of the water. There, at the
far end of the tank, I saw the beads softly floating on the
water. A warm feeling began to flow through me, the
feeling that I was being watched over, and I thanked
Saint Anthony and Ma Ammal, and whoever else might
be up there.

On the long walk back to the station I had time to reflect upon the significance of losing things. It was
strange. You knew they had their tale to tell, but they
were like dreams, leaving hollow gaps and half-remem-
bered spaces. I had noticed that sometimes, after a very
high meditation, I would lose my wallet or my keys, my
earthly identity. They would usually turn up again sometime later, but when you weren't holding the energy correctly something had to jar loose to remind you that
nothing was your own, not even your body, your identity, or your cherished spiritual visions. Staking no claims

72

in this or any other world, you could open unto the
wonder. But the enthusiasm of a Saint Francis could not
appear in its lasting form without a corresponding
strength, an integrity that maintains its own direction as
it passes through the varied forms of this world.

The Vatican

The great dome of Saint Peter's rose over Vatican City
with its carved arches and gargoyles, great halls of gold,
and fine masterpieces of art and sculpture. The riches of
the world had been brought together here and were on
staggering display, presumably proclaiming the supremacy
of the Lord. With the ruins of antiquity translated into
churches, this home of Christendom felt like the glory of
Rome all over again, decked out in a thinly disguised
pagan art and symbolism.

As I walked through these halls, my mental artillery
began to collapse. Walls began to tumble. Who could
deny that it was the mind of men which had invented all
this religion? Perhaps it was spirit-inspired, but as the
Unnamed Light filters through the worlds of form, it
inevitably becomes enmeshed and entangled. The mind
has many sets of clothing, and the fabric tends to turn
heavy and opaque.

Here, gathered in one complex, were thousands of
great works of art and symbols of culture, countless
treasures in high-rising halls of marble with golden
pillars, all inert and static. Their entire accumulation
could not breathe life into that which no longer lived, so
the museum of memory, the pilgrimage of memory, arose
as a compensation, an effort to recapture long gone days
of glory.

I sat down by the corridors of the Vatican and looked
down at my own crumpled body, a wretched palace of
sweat, bones, blood, and bile. Could this body of man be

Vatican
Square

the living Church? The rest was just too secured, too removed in the rigid grandeur of polished marble and gold.

The pious visitors moved through the gallery. I watched them pass and then gather around a swirling Bernini figure. They all had their own mirror, their own "history of ideas," and I, the pilgrim-tourist, was left alone to wonder if I would always be a separate watcher, a hungry-ghost consumer of endless events.

The rains came that day, but the pilgrims continued to file through Vatican City in hope, and the lines held as the rains came down harder.

The ancient Appian Way was muddy. I walked through the mud and the rain to the catacombs. The sky cast dark shadows on the walls lining the road. We went in as a group to visit the underground cemeteries, poking through the empty holes in the ground where buried Christians had been stacked up on top of each other. The earth was cold and damp, and the rains could be heard falling overhead, falling indifferently over the graves of the martyrs.

The rain came down heavily. I sat under an extended roof and watched it fall evenly through the thick, grey sky. And with the rain came a parade of icons, madonnas, transfigurations, liquid forms flowing in time. How the heart sought to touch its vision, but images dissolved, or ignited wars and then dissolved. Could hope ever be pure, or was the eye itself the obscuring cloud? Finally, there were the cleansing rains, unqualified in their activity, washing down the city and leaving a clear absence in the air, brightening the trees and bricks in city walls. The rains fell heavily on Rome, but the lines continued to swell. I went into a cafe for shelter. The cafe was the temple of lost hope, refuge for the coffee-ground mind guarded in its separateness and for the buoyant traveler whose eye was a camera lens. Both had abdicated that profound hope, that hope so deeply conditioned to seek its great reward. And there would be no reward on this pilgrimage, no arrival.

The Ruins of Rome

From the Piazza Venezia one could see the arched stone and broken temples of the Forum, haunting reminders of dreams of grandeur for the city of Rome, of the tides of the Classical and Christian worlds which had converged here and mingled like the slow shift of constellations through the sky.

Ruins were risen above the earth, and more ruins lay below as underground caverns. The Colosseum stood as a monument to the worship of power, the universe of glory built on conquest, the cancer of the mind.

A black cat with a lion-like aura leapt out by the tombs near the cave where Saint Paul had been imprisoned. It stopped and stared, leering at me with witchlike eyes. I commanded it to be gone in the name of Jesus Christ, and after the third time, it left, leaving the ground around the Colosseum quiet and forsaken.

I entered the arena and walked through the stadium stands, looking down through the seats into the underside of the Colosseum with its maze of passageways. Broken stones, half-dug tunnels, buried blood, and rent bones— once around was enough. The air was stifling. The place seemed haunted. You could almost hear the vaporous cries of ghosts in the air. It was enough. Outside, the cat had returned and was prowling around the burial places.

Concentric circles wound around one another. How many persons had been crushed under their weight and buried in the layers of earth? The midday sun shone through the Forum, through the stilled archways of the Colosseum, through the whips and chains of medieval flagellation, through the passing of memory leaving shadows to play against the stone walls.

The Synagogue

In the afternoon, I walked over to Pakistani Airlines to check on my tickets. The man behind the computer ter-

The Synagogue

*The Forum
at Rome*

minal, wearing a white shirt and loosened tie, was very
friendly and asked me if I had plans to stop in Pakistan.
When I told him that I probably would, he began to list
all the places I had to see. "By the way," he asked looking up, "where are you staying in Rome?"

"The way it looks at the moment," I replied, "either in
the train station or in the park."

"No, that is no good, not here in Rome," he said. "The
station is full of thieves, they will pick your pocket in two
minutes."

"I can believe it," I answered

"Wait one minute," he said, "I go make some call." I
heard him speaking in Italian on the phone. He hung up
and turned towards me again. "There is a man at this address." He wrote it down on a slip of yellow paper. "He
is from Israel, a friend of mine. I have told him about
you. Go there and you will find a place to stay."

He pointed out the streets for me to take and we
parted. An Israeli-Pakistani friendship, this was bound to
be interesting. I found the place, a brown three-story
building with a sign hanging outside which said, "Pensione—Third Floor." I took the creaky elevator upstairs.

77

Everything was old, which meant that it would also be cheap. I opened the glass door and walked into a wide and dusty hallway. The place looked deserted. I kept ringing the bell. Finally a teen-ager in baggy dungarees came bouncing through the hall. He said that the owner was out, but that there was plenty of room and staying the evening would be no problem.

As it was Friday night, and the pensione was run by an Israeli, my thoughts turned towards Judaism, another religion that I was half non-brought up in, that I had forgotten completely, and that my thoughts never returned to until years later when I realized that I was lost in this world. This night was the Sabbath, and I remembered passing a synagogue while walking by the river. I abruptly decided to go there and catch the Friday night service.

As I walked along I began to feel a thread tugging at me, one which I hadn't been aware of for a long time. I was filled with thoughts about the endurance of suffering and imagined the synagogue as a place of lamentation where a few old sad-eyed survivors of it all would be slowly droning out some nostalgic service, empty and devastated.

I arrived instead at a large, stately building which was in the process of being renovated, having just been declared a national landmark. I immediately sensed a surging joy. One after another the people entered, happy and smiling, shaking hands, established in warm feelings of community. As the air outside darkened and the vapor rose over the river, the light inside the temple was radiating this sentiment of being gathered together in joy.

The service began. Men were seated in rows facing four different directions, forming a square around the altar. The Rabbi, the Cantor, and others dressed in black robes were up on the podium by the Ark. High up over the stage were images of the Ten Commandments inscribed in Hebrew.

I suddenly remembered sitting as a little boy by my great-grandfather who had come from Poland and went to the synagogue every morning. What I remembered most clearly was that while the prayers were solemnly going on, a group of people congregated in the back saying things like, "Hey, Irving, are you going to play golf this afternoon?" Now, years later, lifetimes later, and back across the ocean, nothing had changed. The old crowd was still sitting in the back, all but ignoring what was going on by the altar. Somehow, it all seemed to fit the picture now. It didn't disrupt the flow.

The prayers, soft lulling mantras, droned on and carried my attention to the tablets of the law as all else began to fade. The law. . . what was it, this revelation on a mountain top which meant more than life itself to a people? As meditation deepened, the air began to answer from the heart essence, declaring that the ultimate love and the ultimate law were one:

> The true embodiment of the law will
> never be known until you learn to
> love the law in its divine perfection.
> The law which stands devoid of love
> is like a body devoid of breath.
> In its true understanding, the law
> radiates outward as the star of David,
> the Beloved, as the manifestation of
> the covenant, the meeting, on the ground
> of time and eternity, between the
> Lord and His people, the meeting which
> appears within, when love dawns,
> all-embracing.

The service drew to a close, and I breathed in the beauty and the power of vast and profound tradition. While walking by the exiting crowds at the door, I tried to shake hands with as many people as I could. That was the blessing. The hours had passed quickly. I remembered the pensione and its Israeli owner, and I walked back quickly.

He turned his head in surprise when I said, "Shalom." When I told him that I was sent by his friend at the airline he invited me to sit down. He sat at the desk, a tough looking man with a barrel chest and thick, curly black hair. The wrinkles on his forehead were pronounced. Rings of cigarette smoke rose from the ashtray on his desk. There were traveler's bags lying around the hallway, brochures, and other remnants of tourist activity. Signs were posted on the walls telling you not to do this or that, typical in cheap places for foreign tourists. He assured me that there was a vacant room where I could spend the night and named a moderate price. I explained that I really couldn't afford to stay in hotels. Finally he said, "Okay, for this night I will not charge you." He then showed me into a room where his children were sitting up watching some American television show from the 1950s. I recognized the one with the baggy dungarees. He was completely absorbed in the television. After continually expressing my fatigue, they finally let me go off to sleep.

The next morning I was up before daylight. After I had showered and dressed, I was surprised to find the innkeeper already up and in the kitchen. He put some rolls and butter on the table, poured some coffee, and motioned for me to sit down. He asked me how I slept, what I thought of Rome, where I came from, and where I was going. I told him that I was following the pilgrims of the Middle Ages, and that I was on my way to Jerusalem, only I planned to go by way of India.

The innkeeper's ashtray was already full of cigarette butts. His face was weathered and tan. Despite a protruding belly, he held himself very tightly, and his still muscular arms bulged from out of his red polo shirt. After a while our conversation dwindled down to nothing, a nothingness enhanced by the silence of the predawn hour. There was a slow tension in the room. Then, half turned towards the air and half towards me, between sips of coffee and drags on his cigarette, the innkeeper spoke, his voice resonating like gravel against the silent background.

"I wore dark glasses to my son's funeral. He was killed in the war." He paused to mash his cigarette into a glass tray as bits of ash sprayed themselves around the table. "I was too ashamed that the people would see that I couldn't cry." He paused again and motioned that I should go on eating. "I've seen too much. I know that life is a struggle. Do you know that? I am an Israeli, but I was born here in Rome. They have given me leave to come to Rome and work because during the last war I lost everything; my son, my business, everything. Now maybe I will be able to feed my family, but if another war starts I will go right back to flying planes. But for now, I am here. I don't know how long it has been, thirty years.

"I was twelve when my family disappeared. There was nothing left at all. The town I lived in was bombed out. I knew that if they found out I was a Jew, they would kill me. So I took all the identification papers from the house and buried them. There was a dead boy lying on the street. I took his papers and ran. I traveled for days. There was no food. In one village I was able to get a job in a factory working day and night, seven days a week, just for food. Then, one day, I saw a worker in the factory who knew our family. He had lived in our town for a time. I kept my face buried in my sleeve all day. I couldn't wait to get my money. When the shift was over, I ran. All day, I had been terrified that he had recognized me and told them my real name. I ran and went to another town."

He took a long drag from his cigarette. "I smoke sixty cigarettes a day, and the doctors say I'm in perfect health. What is it to me? I shot down eight MIGs over the Sinai. I was squadron commander. It was almost the last day of the war. We were by the border when our base was attacked. I saw one of my soldiers lying there afterwards. He was just a boy. His whole face had been shot off. I couldn't even recognize him at first. He was moaning when I came close. He was begging me to kill him. I did. Later, I had his body taken and buried. I didn't want his family or anyone else to have to see that face."

We sat at the table for a while. He asked me what I

was going to India for. I told him about some of the holy
men and the miracles they performed and then showed
him some of Sai Baba's holy ash. "I don't believe," he
said. "How can I? And now, I have a wife and three
children." He sat up, strongly pushing his thick hands
against the table. "No, our greatest danger is to relax. We
must fight. We must fight for our bread and for our children." Then he softened a bit, took some more coffee,
and gazed wistfully out the window. Through the window one could glimpse the beginnings of morning, the
sounds and the light. He asked if I wanted some more
coffee. The street noises were now floating up through
the window.

I went inside and got my things together, thinking
about his story. I could have sworn that I had heard the
same story on a T.V. program a few months before.
When I came out into the hall, the innkeeper shook my
hand and said that if I ever came to Rome again I should
stop by. I gave him a little packet of holy ash, my current calling card. I was surprised at how he took it and
so carefully placed it in a glass jar in one of the kitchen
cupboards. I turned towards him once more as I walked
out the door. What could I say? He had survived.

Miraculous Mary

The streets seemed to follow no particular sequence as
they opened into squares with streaming fountains pre-
sided over by bearded stone gods. Beyond were the outlines of orphic pyramids and sacred temples of vestal
virgins. Soft brown steeples rose upwards as solemn
figurines, carved from their sides, looked down onto the
city. Cycling jets of water circled and splashed about a
string of marble cherubs and then spewed forth from
sculpted animal faces. The stoic senate columns were
shadowed by pantheons of gods and ruins of the temple

of Venus. Goddesses with curved breasts arched backwards, holding their folds of hair with extended hands
and perfectly pointed fingertips. They leaned in twisting
grimace, their stone folded drapery wound around langu-
orous waves of water. The nostalgia for the ancient hung
heavy in the air. In the maze of ever continuous creation,
who could ever hold to a path?

Couples rested in each other's arms on the island banks
of the flowing Tiber. At the end of the island by a concrete bridge, a man slumped against some steps, sprawled
out alone while weeds from the bank came up around
him.

A few raindrops trickled down, and the clouded sky began to darken. Soon the downpour would come. I turned
towards the center of the city as the rains began to fall.
Looking for shelter, I walked into an open church in the
middle of the street. A crack of thunder pulsated through
the sky and the rain began to come down thickly, covering the walkways, streets, and buildings. Clear water slid
down the edges of the rooftops and whirled into the gutters, mixing with the dust and grime of the city and gathering into pools by street corners.

Inside, all was quiet. Others had also come in from the
rain, but the church still seemed vacant. The interior was
aglow, lit up with lights and candles unlike the darkened
Gothic cathedrals to which I had become accustomed. I
read the inscription on the wall. This church to the Miraculous Mary had been founded by a Jewish businessman
who had been visited by the Blessed Mother in a vision.
After his vision, he had become an apostle of her intercession and had paid for the construction of the church.

I sat in one of the pews and gazed up at the altar
where there was a large image of the Blessed Mother as
the Miraculous Mary with fountains of light streaming
from her hands. The rhythmic rain pouring down outside
accentuated the inner silence of the church. I sat in deepened meditation. Mary, who was she, and why could one
feel her presence so strongly? Not only on this trip, but

Miraculous
Mary

throughout my own life, whenever there was a deep need within or a troubling question, I would go to her and could feel an inexpressible presence of grace and reconciliation.

The glow of silence deepened in the room. The entire church became ethereal. It was as if one was sitting in a replica of the same church, only made from luminous, subtle elements. As my eyes remained fixed on her form, I forgot the outside world. Then, the image of Mary began to change form, and as it did I heard thought forms floating down from on high.

I saw a young Mary. She was simply dressed, a humble village Jewess. In a flash of light an angel then appeared telling Mary of her destiny, the Annunciation. While remaining plain and unassuming, Mary accepted the inconceivable, the birth of the holy Christ child through her womb. The voice inside continued:

> This story of Mary is the story of each and every soul. For Mary is none other than thine own self, appearing as the archetype of faith. The Christ child is that which is being born within you, that which is unfathomable and ever new, that pristine self which is unknown, and yet born from within thine own heart.

The image began to change. Mary was seen visiting Elizabeth as John leapt in recognition:

> Dost thou remember when upon first hearing of this wonder, how thou wast afraid to share this new way? Accepting thy quest, thou often wondered if there were to be others who would understand. And then, a path opened before thee, behold, the community of the Spirit.

The image changed again into a Mary looking outward with agonized eyes as Jesus had disappeared from the temple:

> Thy faith grows of its own accord, in its own way, and
> thou must let it lead where it will. Thine own path is
> as unique as thyself, alive in its own nature as it grows
> into its own.

Again the image changed. Now, an elder Mary dressed in
black stood in front of the cross. She held her intensity in
firmness. Tears seemed about to break from her eyes, but
they did not. Instead, the face continually transformed,
becoming more concentrated in its line of determination:

> Mary holds her faith and thus holds the power. There
> will be a time when all will appear desolate, when
> thou shalt be strewn out and alone. It is then that faith
> must hold and transform itself into strength, the
> strength to hold to one's convictions through all condi-
> tions, undaunted and firm.

The image then slowly changed into Mary in all her glory
as the radiant queen of heaven surrounded by rainbows
of luminosity:

> The triumph of Mary is that of faith having gone
> through the void. The immaculate conception is thus,
> the entire process, that miracle which arises from
> within, the birth of the Christic Being.

Over and over gain I heard the words, "I am the Immac-
ulate Conception," and in that moment, I knew that I
was all of this; the inside and the outside, the born and
the dead and the risen.

I do not know how long I remained in the church, but
when I got up the rain had stopped. I walked outside and
back through the city, seeing all journeys as one and the
same. I felt that I could walk anywhere, with anyone else
or alone, at a throne or in the dirtiest street, without distinction. I would remain open, for a child was being
born. The birthing process was never easy, and who
could say how long it would take, but the child was of
the Most High, and once begun, there was no choice but
to remain with it, to see it through to the very end.

Athens

The slight weather changes and green-blue sea waves created their own atmosphere on the boat from Brindisi. I had picked up this ferry after traveling all night by train to the port and sleeping in the station. My head was light and my mind was sensitized from lack of sleep. As the boat headed out through the sea, I could feel the quality of thought change with the air itself. Consciousness, rolling with every wave, took one back through the changing sea to the deep blue of the soul; white marble columns chiseled in time, archways of the classic form, each one an entrance into a new realm of selfhood. White clouds looking down towards the sea rolled gently over hills where initiates in long robes sat on stone benches absorbed in the discourse of wisdom. And beyond the mental wonder at all things, the light quality of peace gently floated on each breath, rising and falling with the sea.

The travelers mingled on the boat deck. The air hung damp with absence of purpose, with no sharp edges to confront the sea waves. A group of blond haired northerners sat by their knapsacks sharing smoke and wine. A couple sat by the cabin door furiously making out. Most folks were lulled asleep, propped against rows of chairs or just lying on the floor. Papers, fruit rinds, peanut shells, and empty bottles littered the deck surface. I contracted into separation and became heavy-eyed as the boat sailed into a stupor.

The hawkers were waiting at the boat station. They carried papers advertising an assortment of cheap hotels and youth hostels. A few of us followed a curly-haired American named Doug who had dropped out of school and was traveling around the world. He worked for the hotels in exchange for room and board. We walked through Athens at two-thirty in the morning. For a buck fifty, he told us, we could sleep on the roof of the Byron, just five minutes off Ommonia Square, with full use of the toilets, showers, and the lounge. He assured us that it was the best deal in town. Besides, at two-thirty, it was the only deal.

Italy and Greece

We walked up these creaking, rusty stairs and out onto the roof. There was very little noise coming from the city. The roof was filled with vagabonds and knapsack people from just about everywhere. Towels, blankets, and clothing hung from the lines which ran across two poles. Some French hippies were still up. They sat together in a corner, smoking, singing, and playing an old battered guitar. It seemed like the Byron filled up at this time when the last train arrivals came in, along with a few straggling voyagers from the streets. Doug and I sat down with the hippies for a while. Everyone else on the roof had gone to sleep. Between songs he told me about Athens and how to get around the city. He said that I didn't have to worry about theft or the like, everything was cool up on the roof. I found an empty space, put my mat down, and fell out.

At some point in the night, for some reason, I opened my eyes and sat up. Directly across from me I saw a silver-blonde-haired woman moving like a goddess, spinning and flowing in and out of the most intricate yoga positions I had ever seen. Her face and hair were like glowing moonstone. A Bhadra Yogini—I wasn't dreaming. I just stared. She went on, oblivious to everything, her arms, neck, and spine flowing into one another, rolling as if she was on bearings, dancing poetry in the night. I was totally absorbed in this sight, but I felt myself falling off to sleep again. I tried to make a mental note of it all and fought to stay awake, but again began to drop back into sleep.

It was light when I awoke. Most of the people were still rolled up in cocoon-like sleeping bags. I scanned the flat roof, looking for that silver-blonde hair, but she was gone. I knew I hadn't been seeing things. My imagination just wasn't that vivid. I looked all over. The roof was filled with cigarette butts and shoulder bags set beside sleeping bodies. A succession of heads with long, stringy hair protruded from the sleepy cocoons. There was nothing else. The morning was cool. Like embers from a night fire, the roof was played out and still.

Athens

Outside, across from a cafe, was an empty square where excavations were going on for a Hotel Dionysus. Three men, sitting down, invited me to join them at their table. We drank clear, dark tea. The early morning was cool and grey. The city, limpid and dull, covered the senses with its grey film. The narrow streets were old and uneven. I didn't understand a word the men were saying, but it didn't matter in the least. I assumed they were working on the excavation. I felt quite at home. The morning movements began to build. A horn blew further up our side of the street. Cars passed, and a motorcycle revved up its engine, sending a flock of pigeons flying off in all directions. A large double-chinned balding man in a dirty blue smock stood up on a crate of bananas and with a crooked iron rod began to crank open the awning for the day's business. Morning grunts circulated around the table along with some nuts and olives, bird calls and racing motors.

I began my first day in Athens, with sensations deadened and poor, by sipping tea and gazing over the construction site. There were collections of pipes and steel girders along with all kinds of ugly machinery. Through wire-screened windows cut into the wooden barriers one could see the excavation pit. The earth falls, and a building rises. The Hotel Dionysus, temple beautiful of mortar, girder, and empty air. I remembered my last visit to Swami Jnanananda, when he told me the story of the construction site. One man said, "We are building a building." The second man said, "The company has put us to work here." The third one said, "A building is under construction." That was the one Jnanananda liked, "A building is under construction," no doer, no actor, just motion.

Back at the Byron, Pericles was making coffee behind the bar. He was big and burly with thick, black, oily hair combed in slices. Pericles, who was very proud of his name, ran the whole operation here. I asked him if there was any place besides the roof where I might leave my things during the day. He pointed over to the corner of the room, and I saw about thirty bags piled one on top of the other into a small mountain of belongings. "The only people who ever steal anything around here are the tourists," he said.

When I walked back out of the Byron, the streets were crowded, stuffed with congestion and aimless commotion. It was like any other big city on the globe, rushing into downtown hell. Only here, if you looked up far enough, there was that magnificent mountain, the Acropolis, rising from the dark web of modern Athens, reflecting the light of the Aegean sun. I made my way around the corners and through the streets which lead towards the Parthenon.

The Acropolis

Fruit stands piled high with purple grapes lined the entrance to the flea market. Throughout the maze of narrow lanes, the shops and booths sold everything from old books and army clothing to an extensive array of modern souvenirs. The most popular of these by far was the celebrated statue of a lustful little nude man. He was on display throughout the bazaar.

As the infested market thinned out, the road turned and curled upward alongside the mountain. There the ruins became visible, and each step upward became an entrance into another sphere. The air felt increasingly rarefied and every breath took on the power of vision.

Rows of crumbled stone lay bleached in the sun of a thousand years. The Olympian frieze had gone the way of the Parthenon of Athena, gone the way of the Citadel, and the glories of Phidias. But still, the columns of the Parthenon stood, noble and firm as the image of Pericles himself.

"He made the city greatest," said Plutarch, "and grew to be superior in power to kings and tyrants...but he did not make his estate a single shilling greater." You could

The Acropolis

breathe in, up here, and let the modern city drop away before the wonder of Periclean Athens. What was this presence and this power that appeared as a blueprint for a world to come? Had Pericles and his associates appeared to plant a seed of possibility, a glimpse of harmony in the very substance of earth?

To minds conditioned by consensus reality, this was a later age, a different age. But with all time ever present, this seed remained as an ideal established within the collective soul. The harmony and structured wisdom of the Acropolis was a vehicle, a reflection of that inner radiance. And all beings, as products of history and as dwellers in eternity, were deeply bound to this project whose manifest patterns would serve as building blocks for the New City.

All beings were bound to its dark side as well. Athens, the harlot of Greece, who underneath her clear, bracing philosophies enslaved millions and hauled tribute from other lands to decorate her walls. The seed was planted with all shades of the ancient heritage, but the harvest belonged to the new, to dwell on the highest possibility, to mold the all-embracing vision from the kernel of the ideal.

And here was possibility, the exuding aura of symmetry, the columns in linear balance, an emblem to that harmony in which the mind of man touches the higher laws and merges with formless geometry, the music of creation.

Far down, the city sprawled outward from the base of the mountain. From the streets, the Acropolis was visible, but half forgotten like a foggy dream. And yet it stood as the integral mountain, forms in stone, reminders of a heritage and of an age gone by, while resting firm, timeless, implacable.

Someone had given me the name of a Demetri Stavanopolis and had said that if I was ever in Athens I should look him up. Now there were hundreds of people under

91

*The Parthenon on
the Acropolis*

this name in the phone book, and the only other thing I remembered about him was that he supposedly worked for some travel agency in Athens. I called all the travel agencies and airlines but nothing came of it. The man at the phone desk shrugged his shoulders.

I had been told that this Demetri fellow worked with the Masters, so I kept asking them to help me find him. Someone must have been on my side because the man at the phone desk wasn't charging me for all the calls I was making. A few hours went by. It was almost five o'clock, which was closing time for the post office. I called another name at random from the list in the phone book. A woman answered the phone. "Demetri who? Stavanopolis?"

"Yes, the one who works in the travel agency. Do you know him by any chance?

The Acropolis

"Travel agency, what travel agency? He hasn't worked there in years. I'm his mother."

We met at Ommonia Square that evening, and he took me through Athens and up to his home near Mount Lycabettus. Demetri worked with both meditation and mathematics which he now taught on a college level. As we walked up the hill, he explained that there were various meridians of occult power which converged between the hills of Athens, making it a truly sacred city. He also spoke of different Masters who had appeared in Athens through the ages.

"The Masters work beyond time and space," he explained. "Nevertheless, either by their own power or through receptive entities positioned in certain areas, they channel energies which feed not only esoteric, but scientific, social, and political currents as well. Whenever a new idea is born in a certain area, the Hierarchy may well be guiding or influencing the particular situation. It is even believed that the great Master who was known in incarnation as Pericles was very much involved in the reestablishment of democracy in Greece."

"During the present period," he continued, "The seventh ray is coming into primary manifestation on the planet. In the past, this ray was known as the energy involved in magic and ritual. What is now being born into the cosmic mind is a new understanding of these energies. Ritual will no longer be thought of as an imposed practical or esoteric order, but rather shall be seen as the very way in which beings live their lives, a way that may come into harmony with the higher laws of evolution. Magic, on the other hand, shall no longer be seen as a wondrous imposition upon the fixed laws of nature, but shall be experienced as the infinite play of the possible, the full potential in the moment-to-moment patterns of changing lives.

"The awareness of such patterns as the subtle vibrational qualities in food, in color, in gems and stones, in numbers and relations of time cycles shall open up an activity which lies beyond conventional causality. The key word of the seventh ray then is 'manifestation.' There will be new possibilities for communications and travel as well as new understandings of the healing arts.

"As patterns of energy are investigated and understood, it will be possible to reestablish their balance within the various levels of the human being. Through such understandings, even the smallest acts may be endowed with the greatest significance. Why, just by glancing at a newspaper, very conscious beings may work on psychic levels with the particular situation they were reading about. Or while stirring a cup of tea, one may project a thought or guide something which is taking place a thousand miles away."

The seven rays are theosophical terms used to designate the manifestation of the cosmic forces in the material sphere. Flowing from the Eternal Logos as the seven lamps burning before the Throne, these rays interpenetrate all physical, astral, and causal worlds, with each ray characterizing a certain type of intelligence, determining one's innate approach to a situation. While the rays are always present, they are said to predominate in cyclical sequence. Humanity is supposedly on the threshold of the seventh ray.

Later, at his home, Demetri expressed the hope that the various spiritual groups in Athens could move towards some form of unity. "As usual," he said, "the ego problems are there. Everyone wants to be the leader of his own group. But we must keep on working to create some pure force that can counteract the tremendous negativity in this area. Unified beings working in the Spirit can actually increase their power exponentially. This is an occult law. So if only a small minority of beings in a certain locale increase their vibrational rate, a stability and harmony can come over the entire area. Unfortunately, as far as I know, this has yet to occur in any significantly large portion of this planet. Yet, it is our most important work at present."

The Acropolis

Later on, Demetri took me into his workroom. He was
of medium height, with dark hair and eyes, and held
himself erect with a combined intensity and dispassion.
He loved his land deeply and spoke of Ancient Greece,
her gods and sages as if they were all in the present. He
explained that the power meridian which flowed through
Athens was linked with the civilizations of Egypt and
Atlantis, and that the same Masters had returned in different forms to watch over the evolution of the area.
From the bookshelves which filled the walls of the room
up to the ceiling, he pulled out a volume of Herodotus
and explained how the great historian had spoken of the
sacred lineage extending from Egypt, through Crete and
Athens. He also mentioned how the great adept Iamblich-
us had placed holy talismans in certain areas of power,
some of which had already been uncovered, while others
were still waiting to be revealed. "As the winds of change
bring the adepts closer to the field of human consciousness, these sacred places shall rise again," he said.
"Shrines will be erected, and the study of the cosmic law
shall be accepted as the principle occupation of man."

Demetri's esoteric words were tempered by his scientific
demeanor. His workroom was filled with books on mathematics, engineering, and computer science as well as the
occult subjects. On a table in the corner of the room were
a few pyramids of different sizes. Their walls and insides
were hollow. Only the outlines were in form. "The shape
of the pyramid," he explained, "has the specific power to
contain and preserve thought forms. One may spend an
entire evening in meditation upon a certain subject, leave
that particular thought form overnight, and return to it
the next day. In this way, its power is preserved and may
even increase to the point where correctly conceived
thought forms can almost instantly manifest on the physical plane. We have a small group that meets here every
week and works on these and similar projects."

When I mentioned that I was thinking of going to
Crete, Demetri got very excited and took out a map to

95

point out specific areas on the island where he believed there were concentrations of power. He told me of some mystical experiences which he and his coworkers had by the hills of Heraklion and insisted that on my next visit, we must all go to Crete together.

Some of his friends and coworkers came over that evening. They sat and discussed their work with enthusiasm. As I watched them, I looked out the window towards the hills of Athens and thought of the great school of Pythagoras where the now fragmented disciplines of mathematics, science, and music were seen as keys of the same instrument playing forth universal alphabets of form, color, and sound, the building blocks of manifestation.

Crete

Before going to India I felt I needed a break from civilization and went over to the Island of Crete to rest, fast, and be alone. I began the fast while hitch-hiking to the southern tip of Greece. After I had waited hours for a ride, a young Arab-looking man in a dust-covered wagon finally stopped. He didn't speak English, but before dropping me off he opened a package in the back of the car, took out two boxes of peanut butter cookies, and gave them to me. I devoured them that night as, again, I couldn't get a ride. After walking for miles, I finally fell asleep in some fields outside a small town.

The fields were cold and damp. The ground was furrowed and coarse. Insects flew everywhere. At one point in the night some dogs came sniffing around and began barking but left when there was no response.

In the sunburst morning, however, the Mediterranean air was glorious. I didn't mind walking another six hours and picking fresh figs off the roadside trees as my assorted insect bites dried out in the sun. A young German couple picked me up in the afternoon and drove me south to the

port, where I waited on the rocks for the boat to Xania,
eating olives.

I slept out on the boat deck. All night the magical
island called, as we approached the zone of Africa and
the source. In my dreams, there were ghosts of gold
metal and Zeus clouds, and in the morning, I saw the
volcanic rock of Crete, Atlantis reborn, as bird omens
went soaring over the coastline.

I walked along the fishy shore of Xania to a highway
and hitched out into the middle of nowhere. On the
beach, a group of jagged cliffs formed an open-mouthed
cave. I set up camp there. Having eaten nothing but the
figs and olives, I decided to recommence my fast, to remain alone and attuned to the elements. The bone rock
earth, wind-breath, and blood-surging sea all mingled
here. After combing the area for dead branches and
sticks, I had a large enough pile of wood for a fire.

I sat in meditation over the fire, but night brought
sleep quickly. After so much travel, dreams became very
astral; deep red robes and old friends changing subway
trains from west- to east-bound stations.

The sun and moon hung together as if on a balance in
the dawn-washed sky. The ancient had arisen. As food diminished, astral winds began to flow through the body,
circulating with every breath. The ocean elementals were
calm and healing. The mosquito bites all over my body
were like the mind itself, a disruptive lash of thinking,
confounded when torn from its elemental origins. Minds
wonder, wander, and whisper in the sand dune hourglass
of time. Like the tides go the causal thumpings of drip
drop raindrop thinking. But the tide turns down. The
outer flight recedes, and the pure mind, the bodhi-citta,
remains unchanged like the clear sky.

Wandering Odysseus, walking on edges, was filled with
trial and ordeal. And all steps are likewise, sunk in moving sands. Nevertheless, the voyage is ordained and writ-

ten in windsong sea rhythms. The free heart seeks not to
alter the consequences of its journey. The embrace of fate
becomes witness to a cosmic play of form as that force
shining through the eyes, through the woven nets of night
and day. The animate elements surging through the rocks
of ancient Crete led one towards the inner recesses where,
pure and calm, lies the ever abounding.

My island companions were the elements themselves. At
dawn I would invoke the earth, air, fire, and water,
chanting their mantras and remaining in their presence,
becoming sensitized to the smell of the fire, to the ionized
sea air, to the tides, and the foam-waves of emotion. The
elementals spoke in a way similar to the "little people,"
but their presence seemed to be more primal, closer to
the depths than external visions. I listened to them,
bathed in them, and washed down with brine, sat with
the rocks feeling their grain and texture, their power and
supportive being. Silence of rock surety, purposeless play
of air and sea, they were never out of place and
enveloped all experience.

After fasting and sitting through the days, I would lie
under the night stars and moonglow, trusting the
elements to take me to their land in sleep and nourish me
with their sense of things. Still life began moving into volcanic visions of Minoa, golden visions breaking into
crown chakra. There was little need for sleep here. I
would sit up at night and do deep pranic breathing over
the fire. The breath circled upwards towards the brain,
blasting back thousands of years to Minoan mountains
and temples.

In this vision, seemingly arbitrary positions were seen
to be archetypal; sitting cross-legged by the fire was a
habit developed through countless lives. Beyond the range
of city thought forms, the old mind fell away, its crowd
of thoughts cleansed from one's aura. In this elemental
mind, the mind of the city with its choking roadways and
buildings was seen as the distortion of ambition and its
heavy ancient monuments were ambition's most deadly
form.

Crete

On the last morning, I sat up on the rocks and said farewell to my elemental friends:

My dear island of Crete by the sea, how your fishermen know you, and how might I someday know you as they? Grapes hang in clusters from your vines and volcanic craters rise from your shores. Your deep blue waters with white capped waves are clear, and one can see down to the rocky bottom. The elementals favor you. Venus rises early in the evening and the moon shines unobstructed in its path across the waters.

This path I must follow, the heart-inspired vision... it shall flow forever with its companions—rock and sea, sun and moon, wind, stars, and fire. How often have I asked how much longer I must endure the sense of separation from them. But when your intimate presence swells into being, I know that I could wait forever. After all you are my father, and every fleck of wind and touch of color announces your presence in a way that no man could announce. Such a presence relieves one of all accumulations, of the need to be anything other than one's self, your eternal child, with every breath your wonder.

And beyond the foaming waters and blue visions, through movements of thought and dream, you are my creator and have endowed me with the heart of a seeker. And thus, anything but the whole of you would be incomplete. For Thou art that perfection of love expressing itself through the elements of thy creation.

3
India

Back in Bombay

I had once thought of India as an ancient land of seers
and sages whose wisdom stood on timeless ground, transmitted through lineages of gurus and disciples. Total submission on the part of the disciple, and total responsibility
on the part of the guru, this was the contract. Certain
psychologists, Indian as well as Western, now spoke of
the Oedipus complex as having never come to India.
There remained, rather, a total dependence on the
mother. Authority was not merely accepted. It was worshipped.

My own fascination with the East began at the age of
twelve while reading Max Muller's translation of the
Upanishads in a bathtub. In spite of the attraction I developed to Indian yoga and all that came with it, I was
cursed with the Western disposition to challenge one's
precursor. I believed authority to be oppressive and born
of fear, and it was an undeniable truth that such authority had been used for centuries to keep a land of villages
enslaved to beliefs and rituals while tribe after tribe, people after people rode into India almost unopposed, taking
what they would.

But things could never be so simple. Along with, or in

*Tonga
drivers*

spite of, the notion of the infallible authority of the
teacher came a relationship unknown to those who cher-
ish their independence above all, a living transmission
which could only arise through the medium of surrender.

The airport had been redone in modern glass and steel,
and there was a different feeling among the porters and
service personnel, a new confidence and self-assuredness.
Soldiers in light brown uniforms, stationed by the gates,
displayed an air of efficiency. When a young boy came
inching up to me with that Indian Uncle Tom smile
which I remembered so well, one of the officers angrily
shooed him away like a fly.

The postal clerk tried to rip me off. When I found him
out, he didn't even bother to play dumb, and I walked
away just hoping that my letter wouldn't get torn to
pieces. The bus ride into Bombay brought back waves of

memory as I looked out the window and saw women in dark blue saris and gold nose rings walking in rhythms of serene ease as they carried pitchers of water on their heads. We drove on through patchquilts of villages and fields of wheat baked in the sun. Mud houses stood grouped under the shade of trees, and men with their oxen slowly dug furrows in the earth.

As we approached Bombay, billboards began to appear. Then came the rows of shanties with roofs constructed of everything from discarded patches of industrial plastics to bicycle tires. Rickshaws, autos, and animals filled the streets, and thick clouds of smoke rose from the industrial zone. The noise was almost enough to drown out the rattling of the bus, which constantly swerved to avoid the cars that came careening by.

Here, however, amidst this spawning chaos of sound and motion, you might find, hidden and tucked away in any corner, a mahatma, a "great souled one," who had realized the truth by direct perception. The holy books insisted that it was not enough to simply visit the temples and sites of pilgrimage. For the holy realm could only be seen through the *sadhu*, the holy person. Therefore, throughout India, one encountered this concept of the self-realized soul. Was it factual? Could a person really know everything? Their followers certainly believed so and brought along scores of miracles to back up their arguments, miracles that weren't just written up in some book, but which were still visible for one who would care to see. But such miracles seemed oblivious to the countless ill-clad and poverty-ridden beggars, and to the chaotic streets on the verge of exploding into madness.

Back in Bombay—it had only been a few years and I didn't think it would faze me, but the streets here hit like a bomb. About the closest thing I'd ever seen to it in the West were the old electric bumper cars in Coney Island; bells ring, lights flash, and everybody bumps into everyone else. After making my way out of the bus terminal, I called Ramesh's office where I thought I might have a place to stay. "Hello," I said, "Is Ramesh in?"

"No."

"Well, is he out?"

"He's just coming."

"Just coming, well when might that be?" There was no answer. "Will he be in this afternoon?" I began to get the idea that perhaps I was asking the wrong kind of question. "Excuse me," I asked, holding myself in, "where is Ramesh?"

"London."

So there would be no place to stay. I plowed back through the terminal going off left tackle, to find out where Westerners could make train reservations. Without this government service, it would be a two- or three-week wait to get on any train. While being shuffled from one line to another, I caught part of a theological debate between a clerk and his stationmaster, the clerk explaining why it was necessary to believe in God. The stationmaster looked through his books, obviously annoyed with the clerk, and booked me on an early morning train to Mathura.

I couldn't worry over not having a place to stay. Bombay lay before me. I stepped out of the station and into the street. Outside the gate, three women in rags were huddled together in what looked like an old basket. Their hands were outstretched. A haggard dog, trotting by, sniffed at a puddle which was right by the basket. Frantic taxis were honking all around, cutting each other off. Billboards were plastered all over advertising raincoats, radios, and new miracle elixirs to cure baldness. The streets were filled with cigarette wrappers and newspaper scraps that had been used to wrap food in and were now lying empty on the ground.

People were everywhere, rows thick, swarming around the station. One man wore a card labeled, "Famous Scientific Palm Reader." Another man lay flat on the ground, his leg swollen to the size of a beach ball. His hand was stretched out. Rows of shops lined the streets with ten or more shops lined together selling the same items. First came the metalware shops, then the stationary shops, the cloth shops, and so on and on. How different was this outer world from the airport whose bureaucrats had close-cropped hair, English accents, and owl-like masks of official imperturbability.

I walked through the streets passing theaters booked for three weeks in advance and lassi stands where the open vessels of sweet yogurt drinks were covered by swarms of flies. My mind flashed back on the flies crowded around the beach ball leg of the man lying by the terminal. Burlap sacks were spread over the streets. Rickshaw *walas* or drivers grinned through their teeth, and corruption burst through its flimsy veneers everywhere.

Then I crossed a sight that stopped me completely. In the middle of the street, as if out of the inferno, a coated black hand rose from the dust and beckoned. The hand had withered down to about the size of a match stick. It was a woman's hand, with an open and pleading palm. In the hunger of that calling hand was the entire third world. I was sickened, and I felt guilty by implication of belonging to another culture. I had seen her hand raised and I thought she was calling to me.

I went over to a street vendor and bought a dozen chappatis. After wrapping them up, I placed them by the woman's side in the middle of the street where she lay. Immediately, a band of young boys who had seen me came running after me on their bare feet which were coal black. Their hands were outstretched, and they all wore torn rags as they came calling, "Sahib, Sahib!" Even as I quickened my pace, they pursued ardently. It was too much.

I quickly cut down an angular side street and ran into a bunch of young men with their shirts off, feverishly lifting weights. They were doing bench presses and curls, and then flexing their muscles as they looked in a mirror. If Bombay was a circus, then each street had its own sideshow. The weight-lifting enthusiasts asked me if I'd like to try my hand, but I moved on. The smell of fire rose from dung-heated stoves. Urine flowed through gutters along the sides of the streets, its smell mingling with the fire. Vehicles rattled and drivers sneered. Bombay of post office cheats and a thousand different hells, I wondered how long it would all hold together.

And yet, it always held together, this miracle of India with a fabric knit of a thread which ran far beyond any logic of Aristotle. Here was a mythic dimension inhabiting a ravaged land where, amidst utter chaos, nothing ever changed.

Darkness was coming, and I still hadn't found a place to stay. The Bombay hotel prices had shot up due to an influx of oil rich Arabian sheiks, and those places which were still low-priced would put dread into anybody. I decided to forget about it and went into a small restaurant near the train station. For about sixty-three cents I was given a sectioned metal plate filled with curried vegetables, dahl, chappatis, rice, and spiced Indian pickles. Over the cashier's counter was a large photo of Bhagavan Nityananda, the great Avadhut, the renounced holy man who wore nothing but a little hankie. Photos of other gurus lined the walls of the restaurants. Was there any place else in the world where you found restaurants with gurus on the walls? As I ate I thought of my old teachers.

Tomorrow I would travel and perhaps see them again. The expectation that had once sent me spinning around the subcontinent in order to have the darshan, the meeting, of the holy person was gone. But there was a respect, a respect for those who had lived their beliefs to such an extent that a true power had developed, one which could flow upon the seeker.

And, like it or not, I was somehow connected to all of this, meshed in a net with little regard for geography or time. For in the living transmission there is an ineffable bond, not of attachment or dependency, but of some unnamable quality which would always find you, whether you were sitting in a temple or a bathtub, and draw you to it. But this bond, like all living transmissions, becomes confused, merged into one's conceptual apparatus. No matter how much I loved India with her madness and spirituality, I knew that I had come back here to reclaim lost parts of my life.

For many, a pilgrimage to India is a search for a mas-

ter. But who is the master, the guru, the guardian? How
can you recognize one? And if you did find a true sage,
then what? What would happen? So many expectations,
like mud-filled rivers—the river had to be cleared, stilled,
so one could gaze in deeply and recapture one's own projections.

I felt better after the meal and went over to the train
station hoping to find a place to sleep. The train was
leaving at six in the morning, and I could already feel the
swaying of the trees and the soft earth of Vrindavan,
Mathura, the birthplace and eternal abode of Sri Krishna.

Even at night the streets were filled. People poured in
and out of the station. I was exhausted and filthy. I
slipped into a first-class waiting room, rolled out a bedroll, and ducked into a bath, dowsing myself with water
from a metal pitcher which I had picked up in a bazaar.
Returning to the waiting room, I sat in meditation on my
bedroll. These are the only train stations I know of where
it was not unusual to take a shower, walk around barefoot in a wraparound cloth, and sit cross-legged on the
floor. No one seemed to notice this except for the conductor. I saw him walking in my direction. I knew that he
knew that I knew that he knew that I didn't have a first-
class ticket. He approached in his short-fitting black uniform with his name printed on a small card by the breast
pocket. He asked me for my ticket in a detached, official
manner. I hesitated for a moment, not knowing how to
play this one out, and then pulled out my pocket photos
of the gurus Shirdi and Satya Sai Baba which I had
picked up in town. The conductor looked at the photos,
nodded, and smiled. He looked at the photos again and
then motioned with his hand saying, "Stay, it's all right."

The train left early in the morning. After the initial
shock of dirt and squalor, I was beginning to enjoy the
living earthiness of it all. India was real, and her wonderful spirit absorbed you into her teeming life. There were
so many people, tribes, languages, and cultures. But it all
worked somehow, even the train reservation system, and

I saw my name printed on the car door for second-class reservations. We passed through rows of shacks, shanties, and villages, through the wheat fields, and out into the plains on the way to Mathura and its holy city Vrindavana.

Vrindavan

The ride was slow and grueling, but with each stop of the sweating black train I felt closer to India and more at ease. The stations were chaotic. Vendors, balancing their trays, strode through the train in waves. *Samosa walas* sold fried vegetable patties. They were followed by the banana *walas* and mango *walas*, and the *chay walas* carried large pots of that thick, sweet tea, boiled with milk and loaded with sugar, which is found all over North India. Children sold drinking water through the windows. Other children came begging, their smudged faces curled down in sullen pouts. They belonged to the resident begging caste of that particular station and rarely took no for an answer. Masses of humanity came pouring into the cars, spilling out of the windows and doors. The train pulled away from the station with people clinging to the window grills and jumping up to ride on the roofs.

A day later, at dusk, the train arrived at Mathura Junction. Mathura is a historic city where, it is said, the brahminical hierarchy absorbed a local cult which worshipped the regional hero named Krishna. The Buddha is said to have preached in Mathura, and its state museum exhibits Gandhara art showing a definite Hellenic influence. Even through the sackings of the Moghul Empire and the British Raj, Mathura had remained a most popular place of pilgrimage. For this site, according to Puranic lore, marked the advent of Sri Krishna who appeared as a manifestation of the god Vishnu in order to relieve the burden of the earth.

India

The city of Vrindavan, where Krishna is said to have spent his celebrated childhood, is about ten miles from Mathura. It is said that, although he had taken birth in a jail cell in Mathura, baby Krishna, by his miraculous powers, had opened the cell doors, put the guards to sleep, and instructed his father, Vasudeva, to carry him by night across the Jamuna River in order to escape the persecutions of King Kamsa. Child Krishna was subsequently raised in Vrindavan by his foster parents, Nanda and Yasoda, who belonged to a cowherding community.

After a string of bus and rickshaw rides, along with the usual haggling over the prices, I found the Jamuna River still steaming from the heat of midday. Wrapping myself in a thin cotton *gumsha* or towel, I entered the running river and let the train karma flow off me. I beat my filthy clothes against the stone steps, rinsed them in the green water, and lay them out by the bathing *ghat* or shore, one of the central gathering places on the river, which was quiet now. The pilgrims had gone. Three young boys with slithering brown bodies dove, one after the other, from a landing which was above the top of the ghat, their faces lit with laughter. They splashed around in the water, undisturbed by the fast current.

I sat by the ghat and felt the changing atmosphere. Tinged clouds slowly darkened over the river. Bird calls echoed over the white sandy banks on the far side of this magical abode known as "Goloka," realm of milk-white surabhi cows and wish-granting trees. But now I was worn out. I couldn't take it all in. There were enclaves by the stone temples surrounding the ghat. I rolled out my sleeping gear. The clothes were already dry but the dirt remained, having become part of the fabric. The river was flowing quickly now, darkly, with occasional patches of white and grey slick from the oil refineries in Delhi. I propped myself up against a red clay stone and looked over towards the end of the ghat to a series of Shiva *lingams*, erect shafts of stone, iconic remnants of another cult from another time. Out past the lingams was the first of the thousands of temples in this city, but night

*Village
friends*

*Bathing Ghat
at Radha Kund*

was falling and, curled up with my blanket of cloth
wrapped around me, I fell into the deep land of Braj
Mandal, the magical circle of Vrindavan.

Vrindavan awakens before sunrise. The pilgrims had
already begun their circumambulation of the area and
were clanging cymbals, chanting, and calling out "Jaya
Sri Radhe." I too, found myself chanting. Even after
years of forgetting all about it, the chants rose spontane-
ously as I walked the street. This energy of *kirtan*, this
song of praise, had a different feeling from the long naves
of the European cathedrals or even from deep yogic meditation. It expanded and touched a space where the divine
"name" always resides, in the depths of the heart. Its constant recitation could awaken the *seva bhav* or eternal
serving mood. This mood, this deep transformation of the
heart and mind, was both the way and the reward, the
very essence of bhakti, the path of devotion.
Lines of pilgrims were already standing by the bathing
ghats. Sadhus sat by open temples with their holy books
and paraphernalia for doing *puja*, worship. The wind
blew the water into expansive ripples, and the pastel sky
was filled with sounds of kirtan. The sound vibrated
through the stone and through the land. It was not the
droning recitation itself which was so captivating, nor the
thousand-year tradition of devotion with its intricate for-
mulas and absurd theologies created to justify it all. It
was none of this. It was the simplicity of that sound as it
entered the heart and opened the heart to the whisper of
Jamuna and the liquid flow of prema, the divine love
which infused the land. Without this whisper, this grace
which could transform the mundane heart, there was
nothing. This sound, this vibration, called gently, to
awaken the mind, to glimpse the depths of Vrindavan.
Vrindavan was this eternal mellow, this *rasa*, this
mood. Old lady hunchbacked beggars stood, their withered bodies leaning against the temple stones, holding out
their aluminum pots for alms while chanting "Radhe,

Radhe." Their *bhav*, the inner sentiment of their ecstacy,
was oblivious to the world, and remained undetected by
the worldly, as they took only the minimal necessities to
keep the body and soul functioning in order to finish off
the last remnants of karma before entering into the eternal abode, Goloka, the real Vrindavan, of which this
three-dimensional manifestation was an incarnation. And
when Vrindavan deepened into the essence of the heart
river, Viraja, which washed away the old crust of the
ego, a new body risen from the Lethe of the Orient
would see the ultimate object of desire: Sri Krishna, effulgent and captivating, simultaneously one and many,
holding the world with his eyes.

The next day, as I was walking down the Raman Reti,
a long dirt road which leads out to the fields and where
it is said that Krishna and his brother Balaram used to go
to tend the cows. I saw, as if in a dream, my old friend
Ray Baba riding by on a rickshaw. We had met some
five years before at a *kshetra*, a place with a free kitchen
where sadhus come for food. Ray had been in India for
well over a decade. He had no address, yet we met
again. Whatever karma is, at that moment I believed in
it.
  Ray Baba was a perennial sadhu and a writer of books.
He had grasped the pulsebeat of the sadhu life like few
Westerners alive. He always had his matted hair tied up
in a Shiva knot and wore the garments of a renunciate.
He had a place for me to stay, and I spent the next few
days listening to the history of various lineages and sects,
stories of the great sadhus, and their disciples. Ray, a natural magnet of information, knew everything about
everyone. We decided to go off to Radha Kund, the holy
village about fifteen miles away, which was the center of
the cult of Gaudiya Vaishnavism, the worship of Radha,
the personification of devotion who is the eternal consort
of Sri Krishna.
  As we started out early, Ray Baba, who had just been

short-changed at three different tea stalls, was musing on how it was that a land that was supposed to be the highest place in the universe seemed to be the lowest as well. "It just goes to show that God is inconceivable," he said.

After getting off by Govardhan Hill—which Krishna is said to have lifted with his little finger to use as a giant umbrella to shield the residents of Vrindavan from the storm of Indra, the wrathful rain god—we took a horse-drawn *tonga* over to Radha Kund and went directly to the Radha Raman Mandir, the temple of Krishna Das Babaji, known as Madrasi Baba, who was the only English-speaking soul in the village. When I had lived in Radha Kund, Krishna Das had taken care of me, providing me with meals and a small cottage, and adopting me as his spiritual son. He had been very much against my returning to the West, saying that life was very short, and the mind was easily disturbed in this troubled Age of Kali, and therefore the most intelligent thing a person could do was to take refuge at Radha Kund and die there, being thus assured of passage into the eternal realm.

Radha Kund

The two *kunds* or lakes are held sacred by devotees because this is the spot where Radha and Krishna used to come and bathe while engaging in their amorous pastimes. Although Radha and Krishna are one and the self-same existence, they manifest as separate entities in order to taste the fullness of divine love, arising out of their eternal activities on the plane of Goloka. This *lila* or divine play is the goal of the sadhus who follow in the footsteps of the great renunciate, Ragunatha Das Goswami. The word *das* originally meant "slave" and was adopted by the devotees of the Lord to indicate their status as eternal servants of the divine.

Radha Madhan-Mohan Temple,
Vrindavan

Das Goswami, as he is called, was one of the foremost
empowered disciples of Sri Krishna Chaitanya, referred to
as *Mahaprabhu* or great master, who is believed to have
been an incarnation of Krishna himself appearing in the
devotional mood of Radha. The *babas*, having abandoned
all worldly ties, come to Radha Kund and live in renunci-
ation and in the practice of intense inner devotion called
bhajan. The essence of bhajan is the hearing and chanting
of the divine names of Radha and Krishna. When the
name is received from a genuine preceptor, its power,
which has been transmitted through a lineage of gurus,
eventually reveals the divine form, abode, and activities
of the deity.

Krishna Das had aged, and he now walked with a
limp. Nevertheless, he greeted us warmly, although one is
always conscious that he is wearing a mask. Like the
other babas, he was dressed in a simple white *dhoti* with
tulasi beads around his neck and twelve clay marks on his
body made from the holy earth of Radha Kund itself,
symbolizing the body as a temple of Vishnu. His brow
was brown and furrowed, and his white hair stuck up in
short bristles. After eating, resting, and talking about old
times, Krishna Das agreed to take us around the kunds,
stopping at the important holy sites and explaining them,
a ritual of his for the last twenty years.

We walked together around the partially paved trail,
weaving our way among temples and tombstones of departed saints. Krishna Das, feeling comfortable with us
neo-Hindus, began to expound upon folklore and philosophy without reserve. The stories he told were neither true
nor imaginative. Rather, they belonged to another realm
of experience known as *Krishna-katha. Katha* literally
means "story," but these stories were regarded by
sadhu-devotees as their very own, to be heard and
savored for their rasa, their ecstatic mood which was
understood by those of sympathetic heart. Through the
hearing of these stories, the knots around the heart would
melt, and one would develop an attraction for the devo-

tional consciousness. As we followed him around the monuments, Krishna Das spoke.

"When Krishna killed the demon Aristasura," he explained, "who had attacked him by assuming the form of a vicious bull, the huge footprint of the bull remained in the ground. Later on, when Krishna wanted to play with Radha, Radha refused since Krishna had committed the most heinous sin of killing a cow.

"Radha informed Krishna that the only way he could atone for such a great sin was to go to all the sacred waters in the universe and bathe in them for purification. Seeing that this would take quite some time, Krishna called on all the sacred rivers in the three worlds. They came offering obeisances and beautiful prayers to the Lord. Krishna then ordered them all to combine their waters, and thus Shyama Kund, the Lake of Krishna, was created. Krishna then took a bath in the lake and was purified of his sin.

"After bathing, Krishna turned to Radha and said teasingly, 'So, let's see what you can do!' Taking the challenge, Radha broke off one of her bangles and began scraping the ground with determined motions, digging right over the area where the hoofprint was made. Seeing this, thousands of *gopis*, the associates of Radha, came to join her. Radha Kund was soon built except for one minor detail; there was no water. Sri Krishna tauntingly offered water from his own lake. Radha refused, claiming that the waters of Krishna's lake were polluted by sin.

"After a lovers' quarrel, Radha refused to speak with Krishna any further. The gopis then began to form a brigade and drew pitchers of water one by one from the Manasarova Lake miles away. Witnessing the great effort and the distress of the thousands of gopis, the sacred rivers of the three worlds again came before Krishna and asked permission to help Radha.

"Hapless Krishna, brooding in separation from his beloved, raised his hands in helplessness and sent them all to pray to Radha, 'For I am nothing without her,' he admitted. When all the rivers appeared before Radha, she too lowered her head in sorrow due to separation from Krishna, and she gave in. Thus, the kund was filled. The gopis proceeded to dig out more earth and joined the two kunds together. As Krishna bathed in Radha Kund, Radha bathed in Shyama Kund. They declared that whoever bathed in these waters in the future would attain the seed of love which Radha and Krishna hold for each other."

I observed Krishna Das as we continued along the stone path. He had undoubtedly told this same story to visitors thousands of times, but still, as he spoke, his body began to quiver with emotion. It is considered bad etiquette for a devotee to manifest overt symptoms of ecstacy, and Krishna Das quickly contained himself and moved onwards in a dutiful manner.

One side of the kund was filled with green waterlilies. On the long concrete wall which ran like a rampart against the back of Shyama Kund, there were congregations of large grey monkeys who seemed sure of their territory. Every so often, looking downward, you would catch sight of the wide green backs of the huge tortoises who lived in the algae-filled kund waters.

We stopped by a well where Krishna Das explained how a saint, while digging for water, had found the tongue of the deity Sri Gopal-Ji and had installed it in a small temple by the well. Another small altar contained a large footprint which was said to be that left by Sri Chaitanya himself on his pilgrimage to Vrindavan.

Along the lakes were several small stucco huts called *bhajan kutirs* where great saints and babas had lived and still live in seclusion doing constant bhajan. Krishna Das told us that his own brother was now living in one of the kutirs. "We are very close," he said. "After death we shall join one another in the eternal service of Srimati Radharani [Radha] and her helpers. But right now I must do this work. All my life I have had this desire to serve, and it is my duty to serve those who come to Radha Kund."

Radha Kund

To be totally absorbed, through bhajan, in the mystical groves of Vrindavan was the supreme situation. In each grove or bower, particular groups of gopis served under one head gopi who was an intimate associate of Srimati Radharani. Upon initiation by the preceptor, one's bower would be revealed, and through intense sadhana or spiritual practice, the soul would come to realize its *swarup*, its true form which existed beyond the gross physical and subtle mental worlds. Through divine grace, this form could enter into the realm of pure love and find its eternal function in the presence of Sri Radha-Krishna.

I asked Krishna Das about the stone Shiva lingams which I had seen standing at the Kasi Ghat. "Lingam means smoke," he said, "and where there is smoke there is fire. The unmanifest Absolute hides in the manifest. When the great sage Shankaracharya came to Vrindavan, even he could not stop uttering the name of Krishna."

My question set off a discussion with Ray Baba on the endless disputes between the various lineages, each claiming authenticity and unique revelation, excluding the others. As this talk continued, some of the Radha Kund babas slowly walked by, their small Bengali bodies with shaven heads draped in faded off-white cloth. They were on their way to a temple *kshetra* to collect their rice and dahl for the day. For the babas who had taken refuge here, there were no more arguments or disputes. This was their way of life, and sooner or later, in one lifetime or another, it would lead them beyond suffering and death into the everlasting bliss of Goloka Vrindavan.

We stayed in a small room by the kund for a few days. Chanting continued around the lakes through the night. The pilgrims came throughout the day. Krishna Das came to visit at least three times a day and kept an eye on our every move, making sure that we would not violate the Vaishnava etiquette of behavior. When we were alone at one point, he asked me what my plans were and reminded me that there was a place reserved for me here.

"Remember," he said, taking hold of my arm, "all this

117

will pass. It is nothing but a dream. The only reality is your own bhajan. This human form of life is very rare, and rarer still is the opportunity to have the association of genuine sadhus. You may think that there are so many circumstances for your being here, but in reality, the only way to get to Radha Kund is by the grace of Radha herself. When I first came here years ago, Krishna took everything away from me: my wife, my family, everything. I was forced to take complete shelter of Radha Kund. That was his mercy. The next time you come, come to stay."

Braj Academy

Bansi Wala Gopal-ji, who dresses in a gold robe and turban, and who it is said was taught to play the flute by Krishna himself, told me that Sripad Baba was in town. Of course, it took him over three hours to tell me. When I first came into the open courtyard where he was swaying and singing by the altar of Radha-Krishna, he motioned for me to sit down and made me a captive audience. He chanted, played on musical instruments, smiled, swayed, and made exaggerated gestures towards the altar with his hands. Beads of sweat drenched his turban and dripped onto his forehead.

When he finally put his flute down for good, a crowd had gathered around. The noon offering had been placed before the deities, and the singing had been describing all the wonderful preparations which the devotee offered to the Lord. Bansi Wala smiled widely, raised his arms high in the air, and called out, "Sri Sri Radha Vrindavan Chandra Ki," and everyone added in unison, "Jaya." Roughly translated this means, "All glories to beautiful Radha, the moon of Vrindavan." Every little temple in Vrindavan has its own set of deities. Although the deities are manifestations of the One Supreme Absolute, each particular aspect has its own personality and its own intimate relationship with the devotee.

Bansi Wala owned this temple in Vrindavan and these were "his deities." He was, in fact, somewhat of a local legend and spent his days rolling around in this semi-intoxicated state of bliss, smiling at everyone and singing to the Lord. It is said that Krishna had taught him how to play the flute, and Radha had given him his life's mission—to always sing the glories of the Lord and to keep a smile on his face. Four years ago, Bansi Wala fell ill and his beautiful face became disfigured. People whispered that it was leprosy. But he kept his smile. Now, he seemed fully recovered. His face, deep and dark against his gold clothing, was aglow with energy.

"You have a good mood," he said to me slowly. "You can attain Krishna." When he could not find the proper English words, he would softly call to Radha, "Radhe-Radhe." Then the words would come. "The whole universe," he slowly said, "has been destroyed, first water, then fire.... But Radha-Krishna," his face broke into a wide smile, "are always here."

After telling me what clothes I should wear, what books I should read, and what languages I should learn in order to attain Krishna, he stopped, looked at me, and asked me who I was and where I had come from. When I told him that I knew Sripad, he lit up again. "Oh! Sripad, Sripad-ji!" The most mysterious ascetic, Sripad Baba was notoriously elusive. So when Bansi Wala Gopal-ji told me that he was in Vrindavan, I felt quite blessed. Bansi Wala told me that he, himself, had been trying to get hold of Sripad-ji for the past week, and that it was impossible. "You see," he explained, "Krishna is in everyone's heart. Krishna see Bansi Wala go here, he send Sripad-ji there. Krishna see Bansi Wala go there, he send Sripad-ji here. But you, he will send to Sripad-ji." Saying that, he gave me a message to take to the holy man. Before leaving, I promised Bansi Wala that I would return to his temple soon. He had given me directions, saying that if I followed them I would find Sripad.

The Academy

The grounds of Braj Academy are on Mathura Road a little way past the Ramakrishna temple and the ashram of Ananda Mayi Ma. The buildings, erected by the Prince of Jaipur for his queens, were donated to the academy by the government. In the last few years the academy had grown into a full-blown institution, enlisting scholars and contributors not only from India but from the West as well. This academy was the brainchild of Sripad Baba-ji.

I found him sitting in an open-air room in front of a makeshift altar of Sri Bankey Bahari-ji, the most popular deity in Vrindavan. He exuded an air of total nonchalance, as if he were both there and not there at the same time. Even if one had seen him before as I had, his features are astounding. At first sight he looks like a Mongolian wild man, wearing nothing at all except a faded yellow cloth wrapped loosely around his waist and folded over. His thick matted hair, which has been growing on its own for twenty years, goes out in all directions. His mustache, curled over his lip on each side, gives him a Fu Man Chu-like appearance, and a large brown forehead frames his wide and gentle eyes. His catlike movements and chipped teeth enhance this combination of phantom fright and gentleness.

I knew he was sitting by the altar so that I would not be able to pay my respects directly to him. Following the Vaishnava etiquette, I offered my obeisances to the deity as he did the same. Inwardly, I was thrilled. I was with Sripad in Vrindavan.

Sri, while meaning "beautiful," is also an epithet for Radha. *Pad* means, "at the feet of," so Sripad was one who remained at the feet of Sri, at the source of bhakti. Here in Vrindavan the deities were rarely called by their popular names, but rather had nicknames which bespoke a familiarity with the Lord which transcends awe and reverence. It is said in the bhakti scriptures that divine love proceeds in stages, each corresponding to a different *rasa* or relationship with God. The beginning stage is awe

and reverence, which leads to either an attitude of peace
and neutrality or servitude. As one grows in love of God
and is cleansed of all mundane associations, this love or
prema, as it is called, develops. Out of His desire to exchange love with His devotees, the Lord takes a form and
appears before the devotee in a certain aspect which
draws forth the more familiar relationships of friendship,
parenthood, and conjugal love. In these stages, under the
influence of *Yoga Maya*, the energy of divine play or
Lila—as opposed to the regular *Maya*, which indicated
the illusory forms of the temporal world—the devotee
completely forgets the distinction between himself and the
Lord. He forgets all about the concept of "God," and just
sees Krishna, for example, as a beautiful little boy with
soft, cloud-like skin and lotus eyes. However, it is cau-
tioned that such relationships with the divine should not
be misunderstood. They are enacted on the plane of eternity as the *Nitya-Lila*, and are in fact the archetypal
origins of all mundane relationships.

There was never any such thing as small talk with
Sripad. I just sat there with him in silence. Eventually,
speaking very softly as if I'd been away for five weeks
and not five years, he asked me how long I planned to be
in Vrindavan. As I began to answer, I instinctively pulled
back, suspecting a trap. I gave all sorts of reasons why I
could not stay for any length of time, saying that I had
no money, my plane ticket would expire, I was needed by
my family, I had work to finish, and so on. "Do not
worry," he said easily, "the Lord will take care." I knew
right away what this meant. Baba wanted me to do
something for the Academy. It was one of his well-known
moves. But who would want to say no? People were coming from everywhere, falling all over each other just to
get a look at him. If you actually got to see him, you
knew that something was happening. He called out to
some of the people in the compound and told them to
prepare a room for me and to arrange for my food.

It turned out that Baba wanted me to help edit the
English edition of the Academy journal which was to be

sent to the printers in two weeks. At least half of the articles had not arrived yet, but this didn't matter. I was more than glad to get the job. This would be my opportunity to be in the constant presence of this most renowned holy man, whose vision to create a temple of higher learning was coming to fruition. I hauled my things up to a small room, where I was given a desk and a stack of papers to begin wading through.

Braj Academy had attracted the attention not only of the Vrindavan community but of the entire country. The Academy had already done much in its effort to preserve the Braj heritage by collecting books, ancient manuscripts, paintings, and sculpture, as well as sponsoring historical studies and archeological digs around the area. As I sat with Baba the next few days, he explained the Academy to me. It was based on the ancient academies of learning, as the open-air academy of Plato, where seekers of truth and wisdom would come to exchange ideas and to learn. Its particular function was to preserve and expound the Braj heritage as well as that of the entire land of India. Sripad Baba made it clear to me that he was not only speaking of external academic information, but also of the deep sadhanas and esoteric teachings which had been given by the Masters and Acharyas who understood the full import of the *Vedas* and *Upanishads*.

The Academy itself was a large walled structure with a great open-air courtyard in the middle. There were elegant pillars placed around the yard and palatial designs on the upper walls. We would sit in the afternoon taking tea in metal cups, Indian chay, with milk, sugar, and dried tulasi leaves from the altar thrown in as *prasad*, consecrated remnants from the offering that had been made to the deity. There would always be visitors, some of whom came from quite far away, but the atmosphere remained relaxed and tranquil.

From within the courtyard you could see the amazing twisting trees of Vrindavan hanging over the walls. In the early mornings, the peacocks came out and paraded on the roof. Baba explained how the cultural heritage of India, that spirituality which had existed as a timeless fabric throughout the upheavals of history, was in danger of slipping away. The educational system imparted by the British had been designed to reduce India to a nation of clerks. In order to resurrect the Indian mentality, learning would have to be united with values of devotion and wisdom.

"No matter what one may think or believe," Sripad Baba said one afternoon, "you must come to understand the special nature of this land." He spoke softly, as if this was a simple fact which I would have to comprehend sooner or later. "This land," he continued, "has been graced by the feet of the Lord. His appearance has sanctified it forever. It has withstood invasions and persecution. Even after it has been leveled to the ground, it has sprung up again as a center of devotion and learning."

This led to a discussion of centers of power and the nature of holy places. The holy land or *Dham* was nondifferent from the deity, being an incarnation of the deity's own attributes. While the holy abode was eternal, it descended along with the incarnation of God to bestow grace upon the fallen world. Vrindavan, then, was nondifferent from Krishna, as the holy mountain of Arunachala at Tiruvannamalai was itself a form of Lord Shiva. Sripad-ji went on to explain that every *thirtha* or sacred place had at least one living saint there in order to "hold the power."

Ray Baba also came by, and Sripad put us both to work, editing during the day and typing all sorts of letters at night. We often stayed up late at night with Sripad, and he would talk freely. One night he spoke of the different avatars of Vishnu as mirroring the progressive development of consciousness as in the gestation of a child in the womb. Each avatar was a further unfoldment, incarnating another quality of the Absolute Consciousness. The incarnation of Krishna was seen as a complete manifestation of God's love, transcending all conventional bonds, while Rama, the previous incarnation, brought down with him the quality of dharma or pure righteousness, which Sripad saw as inferior to that love which appeared as non-dharmic to the untrained eye.

Ray informed me that it was very rare to find Baba in such a talkative mood and that we should take advantage of it while it lasted. When I asked Sripad Baba-ji about the difference between the worship of Vishnu and that of Shiva, he laughed at my imagined duality. He said nothing else about it until an hour later when he turned and softly stated, "Shiva *is* the turning inward of awareness into the non-dual nature of reality."

Never perturbed or disturbed, Baba would answer questions when he felt like it. At other times he would remain silent. Sometimes he spoke in such a soft voice that he appeared to be shy like a small child, and you had to strain to hear him. At other moments, he would yell at someone, taking on a fearsome, tiger-like aspect. His presence was so intense that your thinking process would come to a halt when you were near him. I found myself feeling very awkward.

It was not as though the answers Baba gave were something which I had never heard before. In fact, after the first week, there was nothing more to say. What was fascinating was the way in which Baba answered the questions put to him. A visitor would come and ask some serious question or ask for advice on some difficulty, and Sripad-ji would just sit there in total silence. Sometimes he sat for a good half-hour, oblivious to everything, not caring if the questioner remained or walked away in disgust. It was as if he had gone somewhere else to get the answer. And then, out of nowhere, he would answer, quite matter-of-factly. This is not to say that you would receive a straight answer; just about everything that Baba said and did was rather inscrutable.

Early in the morning, for example, a long black limousine would appear in front of the Academy with two or three dignitaries dressed in either Western suits or clean, white brahminical dhotis. Baba, walking up barefoot and ragged, would be ushered into the front seat and whisked away. Three days later he would return at three o'clock in the morning and then would not be seen again for another two days. Suddenly he would appear at the door asking you to come and type some letters.

I spent a good deal of time typing Baba's letters, which he always signed "Yours in the Lord." Sripad-ji would stay up in the little office room until well past midnight, dictating and inspecting everything. The electricity was constantly going out, so we would set up candles to see by and would endure the heat and the mosquitoes. Around one o'clock in the morning I would get so tired that I would begin to doze off and start making all kinds of mistakes. Sripad would send me off to take rest and Ray Baba would come in and take over. When I awoke in the morning, the typewriter would still be going, and Sripad would be sitting there as if he had just begun while Ray would be falling into a stupor. So we would switch again.

This was most menial work for a holy sadhu such as Sripad who was literally worshipped by the people here. He would work through another day and night, completely calm, with no visible trace of anxiety or fatigue. I had heard of Baba's work habits from others. He would go on and on, pushing as long as he could for days at a time. Then he would disappear and sleep for a day or so. In this way he engaged the maximum amount of time and energy in *Bhagavan Seva*, the service of the supreme.

We worked through the days and in the evenings went to visit the temples and walked through the bazaars. Sometimes I would go out to the Raman Reti very early, just at dawn, or in the evening at twilight. Peacocks would strut in the fields fanning out their feathers. Down by the Jamuna's sandy banks, buffalo would slowly saunter, their sludgy brown bodies moving over the white sands in placid ease. The roads were dry and powdery beige. Temples curved along the riverside, bounded by crooked trees in contorted dance positions. The beggars lined up by the temples. Hairy, mud-stained pigs picked the garbage from the open gutters which lined the streets.

India

I walked past the tree from which Krishna is said to have jumped when he chastised the serpent Kaliya. Certain Indologists now believe that the Kaliya serpent was really a metaphor for a sect of Naga Babas, tribal snake or phallic worshippers who were supplanted by the Vishnuite sects in the area—with the converging of myths serving to unite the two communities. But the fields with their stalks of white flowers, the groves of dancing trees curled around one another, and the aged temples of smooth stone held their own answers.

In the evenings, the temple of Behari-ji was crowded like the central station. Pilgrims, devotees, visitors, and the *brajabhasis*, the residents of Vrindavan, all mulled around the black-and-white stone floors waiting for the curtain to open. Then the gongs began to ring and the crowds closed in around the altar. The curtain was parted revealing the figure of Behari, a small black deity, and everyone went wild. Some pilgrims sang, others were jumping up and down, others had their arms held high in the air. A priest made offerings of incense, *ghee* or butter lamps, and flowers to Behari-ji and brought the offered articles over to the edge of the platform where people touched the burnt-out wicks to their heads. Everyone threw coins up onto the altar. Some threw bills and were given little clay cups of prasad by the attending priest. Even after the curtains closed and the excitement leveled off, many remained in the large open temple square. Large pots of sacred *tulasi* plants were positioned in each of the four temple corners, and the temple-goers circumambulated around them, bowing and touching their heads to the pots.

Vrindavan, it is said, was named after the gopi Vrinda, who was very dear to Lord Krishna. During a lover's argument, Vrinda had put a curse on Krishna to become a stone. Lord Krishna, in return, cursed Vrinda to become a plant. Lord Krishna then became the sacred *salagram* stone which is worshipped on altars along with the deities, and Vrinda became *tulasi devi*, who grows wherever there is pure devotion to Krishna. Here in the temple, tulasi was growing profusely, her leaves spreading in all directions.

One man, his emaciated body smeared with ashes, was standing in the corner on one foot. His eyes were closed, and his hands were locked in a difficult yoga mudra. Passing pilgrims threw coins at his feet. The pilgrims came from all over, especially during the holy autumn and summer months of *Kartikka* and *Sravana* as anyone who lived in Vrindavan for these months was freed from past accumulation of sins. There were ashrams which took in thousands of pilgrims during this time, and many of the temples served free meals during the day which, in the holy months, became feasts of celebration.

The priests, a little on the fat side, sat by the altar and guarded the deity. In the back, on a slightly raised platform, musicians sang *bhajans*, playing harmoniums, cymbals, and clay *mirdanga* drums. Outside, a stage was being set up for a *Ram-Lila* theater performance. These lilas presented in song and dance the divine stories of Rama and Krishna. Thousands came to these spectacles. During the performance, those in the audience would identify the actors with the deity himself. Very often, at the end of the production, they would carry the leading actor home, not allowing his feet to touch the ground. The actors came from a family lineage and had been specially trained since early childhood. Their careers, however, were short-lived. By the age of fourteen, one was considered too old for the roles and was forced into retirement. I knew one such actor who, now in his twenties, was still the celebrity of his village.

The entire atmosphere, both inside and outside the temple, was one of carnival gaiety. It took only a short stay in Braj to make one realize that the carnival was, in fact, perpetual. Rarely did you see anyone in a somber mood. The residents of Vrindavan always seemed to be enjoying themselves in a twinkling way. Perhaps this is why the bhakti theoreticians had written that if one can-

not perform any rigorous sadhanas or devotional pro-
cesses, one should simply come and take shelter of
Vrindavan. That would be enough. The power of the holy *dham* would act on anyone residing within its borders,
awakening the serving mood, the bhakti rasa.

Before leaving Vrindavan, I went with Ray Baba up
Mathura Road, past the camps of the Ram Bhaktas who
were sitting, blissed out, smoking *biris* and other assorted
paraphernalia. We went over to the *samadhi*, the
memorial tomb, of Neem Karoli Baba Maharaj-ji. The
shrine was surrounded by two temples, one with a large
image of Durga riding on her tiger, and one with the image of Hanuman, the celebrated monkey servant of Lord
Rama. The energy by the shrine was tremendous. It
poured out of Maharaji's picture which was installed in
the unfinished dome of the samadhi. Some of the ashram
devotees allowed us to sit in his room by the bed, where
he used to lie, covered only by his old woolen blanket,
and conduct his lila. The walls were lined with written
Ram mantras. The rest of the room was filled with a very
real presence. We sat and bathed in it. Maharaj-ji was
laughing.

In the evening we walked over to the Ramakrishna
Mandir and sat for kirtans. Then at eight o'clock we
made our way along the wide dirt road, passing the
Ramakrishna Mission Hospital, and came to the ashram
of one of India's most revered saints, Ananda Mayi Ma.
    Whenever Mother came to town it was a great event.
The crowds had already gathered outside the two-story
brick building where Ma was staying. There was an intense atmosphere of expectancy. I took it as a sign of ben-
ediction that on my visit to Vrindavan I would be so
fortunate as to have the darshan of Ananda Mayi Ma.
    It is said that Mother, as her disciples call her, did not
receive formal initiation from any guru. Rather, at a very

early age, divine phenomena began to manifest in her body and behavior. At first her family feared that she might be possessed by evil spirits and sought the help of healers and holy men. All of them, however, were amazed, as she was manifesting symptoms similar to those Chaitanya himself had shown long ago. He, too, was thought to be a madman by the uninitiated. Her visions of God continued, and ultimately her husband Bholanath accepted her as his guru. Over the years, the various stages and states of divine ecstacy progressed, and thousands came for her darshan and blessing. Her teaching was as universal as her being. She spoke of the "pathless path" and met all according to their needs.

At a precise moment, the door was opened, and we were led up to the roof and sat. There was total silence. Chanting softly echoed from the many temples in the holy city. Without a word, Mother rose from the cot where she had been lying and sat in silence and serenity. Her physical body was well into the eighties, but her being appeared to expand like the full moon, bathing the area in an indescribable coolness. Bliss emanated from her being, along with a wholeness, a complete sense of ease. The entire crowd on the roof was taken up in meditation, enrapt in this effulgent presence.

After some time, the temple bells rang and a soft kirtan began. Mother took some prasad, little chips of sweetened white flour, and threw it out among the crowd. But it was her presence which told all, the ever-outward radiance of serenity and completeness. I felt so light, I was not aware of the floor beneath me. To gaze up at this form and see the Absolute embodied before my own eyes is to understand the position of the guru. Here, there was no longer a human presence but something entirely otherworldly.

But the form was human, and here was proof of the path, proof that it all wasn't wishful imagination, proof of the possibility of human evolution. In this quietude, one saw the end of human attainment, the depths of a peace that could never pass. Here was Vrindavan Chandra shining in the night, and here, behind Mother India teeming in confusion, was the Great Mother, implacable, inexpressibly aware, filled with grace, shining past the forms of time into the everlasting.

Entering the Himalayas

Ray Baba saw me off at the Vrindavan bus station. We arrived at daybreak while the cows, dogs, and other street animals were still lying asleep on the sides of the road. But smoke was already rising from the tea stalls, and by the time we arrived, the station was filled with people.

A group of men surrounding a blue tin bus were making a concerted effort to push it out of the station and get it started. As the bus approached the outer gate, a man on a bicycle came careening by a little too close to the tea stall, and knocked a woman and her child into a watery ditch. They came up muddy and screaming. As a crowd gathered round, the bus engine finally caught and began to sputter. It rumbled out of the station, followed by a mass of frantic passengers trying to pile inside.

We sat on a wooden bench by the tea stall with chay and rusks, a kind of long toast that Ray put back on the fire. Being low caste foreigners, we did not merit throwaway clay cups, but were given our tea in old stained glasses. "Perhaps we are the beginning of a new caste," Ray said, "neither Americans nor Hindus. I will have to find a good name for it." Ray, who knew all the ropes around the bus terminal, was able to get me a seat on the bus going to Delhi. Once in Delhi, however, I was completely overwhelmed and submerged in a moving, grasping morass of humanity. I tried to struggle and push my way out of one bus onto another. I spent the night sleeping in the aisle of a bus, caved in underneath hosts of bodies, sinking into metal and grime. The bus bounced along to Hardwar and did not arrive until next afternoon.

The holy city of Hardwar is a great site of pilgrimage. It was here that the nectar of the heavenly Gods is said to have fallen on the Ganges. The Ganges, in fact, had been diverted a way out of town, and the pastel bathing ghats and large statues of various deities lined a great canal. But it was here that the pilgrims came on tour, taking holy baths, walking along the river, and buying all kinds of items and holy trinkets from the shops. One usually took a flask of Ganges water home and kept it until next year's pilgrimage.

I took another bus up to Rishikesh and began to walk up towards the Shivananda Ashram. Hermann Hesse visited this thirtha on his journey to the East not so long ago. Thousands had followed. On one side of the road were the "Swiss cottages" where a certain swami catered to Western nomads who sat around smoking marijuana through most of the day. On the other side were the tourist bungalows. It was difficult for Westerners to gain admittance into the ashrams now. In the past, they had violated so many rules and had made such a mockery of the holy customs that you either had to know someone, or somehow prove that you were a genuine and serious aspirant in order to be allowed entry.

If you cut to the right about a half mile before the Shivananda Ashram, you could find groupings of white rocks which led to the bank of the Ganges. A few buses were parked near the rocks waiting to be washed down. Children were playing, and solitary sadhus sat by a Shiva temple about a hundred yards away. I walked out to the rocks and bathed in the cool, green mineral waters of the Ganges. Ma Ganga, flowing down from the mountains, was electric and vitalizing.

A few miles further up the winding road was the Lakshman-Jhula bridge. Lines of beggars calling for *bakshish* or alms, waiting taxis calling for passengers, and assorted shops calling for customers surrounded the pilgrims who crossed the narrow span which wobbled

131

with every foot that hit it. I found my old sitting place a way up and sat cross-legged on the sun-bleached rock. The Himalayan hills rose from Ganga side like immovable slopes of infinity. A gentle breeze blew the trees on the mountains, which were dotted with red flags from the *kutirs* or huts of hermits. Ganga, jungle, and Himalaya all rose in such power, with the surging river stretching over the rocks—endless heights, and endless skies.

Before leaving I took a flask of Ganga *jal*, the holy Ganges water which millions of pilgrims bathed in for healing and purification. The water was said to flow from the Karana Jala, the causal ocean, through a hole in the outer edge of the universe that was created by the big toe of Vishnu.

Other teachings spoke of the Ganges water as flowing down upon the head of Lord Shiva before cascading through the mountains. Following their Lord, the Shaivites claimed the Himalayas as their own and sat with their ritual tridents and ashen colored bodies on the banks of holy rivers, by waterfalls, and in secluded caves.

Across the bridge, about a mile south and a few hundred feet up in the Himalayan foothills was Tatwala's cave. Tatwala Baba, a powerful ascetic who had lived and reigned on the Ganges, had been shot and killed in a bizarre incident some years ago. His cave was now a monument. A few of his disciples were still there, but most of them had moved on. I sat for a while and listened to a sadhu assure me that Tatwala Baba was now at Gangotri, the source of the Ganges, performing austerities.

Wandering by Ganga side, I passed a number of ashrams where I had previously lived. I lingered by the gates, but I couldn't bring myself to go inside. Some memories were better left alone. But there was another realm, a realm in which the ashram, the holy thirtha stands alone, out of the world, alone at the *sandhya* of twilight and dawn, the junction where imagination and memory meet the never-ending open.

Swami Jnanananda

I survived the bus ride through the mountains and, after walking up a series of trails, found the familiar dirt pathway which turned off to Barlow Ganj. Here the mountains rose to their heights, their faces filled with forest growth. The air was perennially cool as the mountain walls blocked out the sun. With each step the rattling bus vibrations fell away. Here, in the Himalayas, was renewal.

I clearly recognized the area now, and my heart began to pound. I wondered if Swami would even be there, or if he would be off on one of his journeys into the upper mountains. On the final curve, by the waterfall, was an old abandoned distillery built by the British, who once used these swift running cascades to make beer. As the road descended, I could peer down into the ganj, and I thought I saw him. Yes, it must be he! He was unmistakable, with that orange cloth and mane of white hair, a walking stick in his hand. I moved my aching, tired body as fast as I could and finally came to the wooden flat and winding gravel road. Pink and blue flowers bloomed wild all around the cottage which was known as "the House of God." I stopped and called out loudly, " AAAOOO-UUUMMM." He looked up. I ran towards him in love and fell at his feet. I was back with Jnanananda in his Himalayan home.

He motioned me over to his small kutir which was on the other side of the cottage, and we sat. Serene and undisturbed, he picked up a garland of flowers that he had been weaving as an offering. He wove slowly and precisely, sitting on a mat of woven pine needles with his legs crossed and a faded orange dhoti folded about him. Nothing had changed. Five years was not even a minute. He looked exactly the same, sounded the same, and even wore the same clothes. Suddenly, the last five years of my life vanished. Here, in the coolness of Himalaya there was no passing of time. His white skin and white hair seemed to merge with the air. Swami spoke, "It's good that you have come. I have just been sitting here doing my sadhana." He laughed and went back to his weaving. The air was quiet, and the trees and mountains shaded us from the sun. After sitting for a while, Swami sent me off to wash up and rest.

Later that afternoon, feeling relaxed and vibrant, I walked up the wooded trail to his cabin, which was covered with foliage and set in the rising wall of the mountain. There was a stone walkway leading to the entrance and a roof held up by two poles. All around the cabin were brilliant pansy-like wild flowers. Past the flowers was a level grassy area with a clothesline and a place for washing pots and utensils. A bit further away was a deep *havan kund*, a symmetrical cow dung pit which was used for ritual fires. About thirty feet from the cabin was the circular kutir, about six feet in diameter, made entirely of straw and cow dung.

Swami had a guest inside, a young Indian who had come for darshan. He motioned for me to come in and sit down. Swami asked the man how long he was able to maintain a full lotus position. The young man, wearing Western clothes, crossed both his legs to demonstrate the posture, straightening his back. "No one can sit in this posture for more than an hour unless he has done so in a previous existence," Swami declared. He encouraged the young man to continue his practice of meditation, explaining that his own guru had said that the desire to meditate is itself the blessing of the Lord.

Swami continued speaking. "Along with meditation there must be bhakti. It is essential...and where does bhakti come from?" There was a lull in the cabin. "One is born with it! That is why," Swami said emphatically, "the yogi does not seek out others to teach. Only those who have the *samskaras* (latent impressions in the mind from past lives) are able to hear, and they will appear out of their own accord."

When the visitor had gone, Swami prepared his late meal, and we ate together in the kutir which had already

withstood three rainy seasons. Jnanananda asked me
about my activities over the past years, and I spoke of
various teachers of meditation and healers with whom I
had worked as well as this present pilgrimage. He seemed
amused by it all, and it did feel rather paltry in the rarefied air of these mountains. The meal hadn't changed
either—a light mixture of rice and dahl sprinkled with
ghee and a few vegetables.

After we had finished and cleaned up, we remained in
the kutir. Swami, leaning with his back to the wall,
gestured out towards the Himalayan air and spoke.
"Never join any organization. One may, of course, work
with people, but always remain free." There was another
lull, like the open spaces between the rising ranges.
"There are certain basic principles," he went on, "which
are necessary if one really wants to live a spiritual life. If
you do it, do it completely. Otherwise, you might as well
go back to the world." Swami sat effortlessly, not suppor-
ting himself at all. He was so light that he could have
been blown away, and yet he was firm. "Always be free.
Live in that reality which is beyond time. Never own
anything, and never work for a living."

"But Swami," I cut in, "you must understand the ways
of Western society. There are no free kitchens for sadhus.
One must do some kind of work." Swami remained silent.

I had first met Jnanananda in these mountains some
years before. I felt close to him in a way that was timeless and thought of remaining with him in the mountains.
But I knew that I had to come down and live my life,
and he knew it too. In meditations, however, I was able
to close my eyes and feel the presence of his cabin, as if I
were right there.

Swami Jnanananda roamed the mountains with his
staff, like a lion. He was free. He was gentle, and he was
a friend to all. The merchants and hill people waved
hello to him. The sadhus saluted him, and he was not
even an Indian!

Swami had been born in the mountains of Switzerland.
One day, in his teens, he attended a show given by a

psychic reader. The psychic had read the future of some people whose names had been picked out of a hat. He had picked Swami's name, but instead of giving a reading he simply walked over to the table where the young man was sitting and said, "All I can tell you is that in a very short time your life will completely change." A short while later, Swami found the works of Paramahansa Yogananda. He wrote to his ashram in Calcutta saying that he was deeply moved by the teachings he had read but felt he had to draw directly from their source. A short while later, he went off to India. He took nothing with him but the clothes on his back, and an umbrella which he had borrowed from a friend. Upon his arrival in India, he mailed the umbrella back. Meanwhile, his mother, disturbed over her son's departure, wrote a letter to Dr. Carl Jung asking for advice on how to handle the situation. Jung replied that she shouldn't worry, it was just a passing phase. Swami never returned.

We sat through the afternoon. The winds stirred the flowers and grasses as the Himalayan elementals held us in their cradle of power. Even during the spring and summer thaws, the mountains held their aura of cold strength. For this was the land of Shiva in his continuous posture of meditation.

"All rituals belong to the outer world," Swami explained. "God, puja, and the rest...the inner life is silence. Meditation may not yield amazing experiences, but it will put you in a receptive state through which intuition can flow. There was a young boy here about a week ago. He was the son of an Indian devotee, and he was very perceptive. He said to me, 'Swami, everyone is trying to be somebody.' Dropping his voice, he turned directly towards me and looked into my eyes. 'So, I say be a nobody, no-body, renounce all bodies; gross, causal, astral, until you stand completely alone.' "

Later on in his cabin, Swami asked me about Braj Academy. I went on praising the open-aired academy modeled after that of Plato, the projects to reconstruct the Braj heritage, and the collecting of ancient manuscripts and works of art.

Swami looked up and laughed. "What is in danger of falling down, can it be worth building up?" He laughed again, lovingly, and then spoke. "Organization leads to disorganization. It is not that I am opposed to anything, but for a yogi any work that is done externally can better be done internally. Sripad Baba-ji is a great mahatma. He could be sitting in a cave anywhere. He has no need for any Academy. It is all just his lila."

The sun was beginning to dissolve into colors which trailed through the clouds. I looked at this wonderful man in the faded orange robe with his long white beard. He didn't have a care in the universe. Sitting firmly on his mat, he chanted a long AAUUMM just as the gong sounded from the Christian Brother's School, a long way off in Mussoorie. I followed him into meditation. Thoughts began to dissipate into silence. The room filled with energy, no form, just the stillness and vast inner space. Jnanananda did not use the word "meditation" often. Instead, he called it "sitting in *jnana*," sitting in knowledge.

I was up early the next morning. The Himalayan water was ice cold. I doused myself with two bucketfuls and sang "Ananta Hari Om," the way Swami had sung it after meditation the night before. By the time I walked up to Swami's cabin, the rays of the sun had begun to shine through the mountains, warming the cold starkness and mellowing the winds. He was inside, sitting still, sitting in jnana. Bird calls carried long through the mountain air.

I scanned the room. Every object was in its right place. The ceiling, painted gold, had a large yantra drawn into it. Pictures of various deities hung from the walls, which pulsated with light. In the far corner was an image of the deity Bhadra-Narayana, whose temple lay hidden among the snows of Badrinath. Around the deity were other pictures which I took to be Swami's lineage of gurus. A grandstand of gods seemed to be stationed around the room. On another wall was a collage of many different saints and sadhus. Some, such as Sripad-ji and Ananda Mayi Ma, I recognized. Others I did not. A ghee lamp was softly flickering in the corner by the altar, and some fragrant *dhoop* incense rolled up on a string was burning down as it hung from the shelf-like ledge.

Later that morning we ate *parathas* (fried bread) and drank tea. As the steam from the hot water rose in the cabin, Swami showed me how to make parathas with one hand, folding the dough with elf-like spryness. The smoke from the ghee lamp merged with the incense, the steam, and the air. Outside were the lively noises of morning. Swami's eyes glowed as he sat, wrapped up in a woolen blanket.

I had been surprised to see a picture of Sai Baba among the assemblage of saints and sages on the wall. I asked him about it, and he told me that on a recent pilgrimage to South India he had stopped at Sai Baba's ashram for two days. Baba had come over to him and smiled. "It was a very special smile," he explained. "My own guru used to smile at me in exactly the same way. No one else could have known that." He gave me the address of one of the ashram administrators, telling me it would be useful to have when I went south.

We walked down the mountain that morning to visit a family who had asked Swami to come and see their sick father. Swami walked swiftly down the trails. He carried a walking stick in his right hand, which he planted firmly on the ground between strides. As we walked, Swami told me about a 125-year-old Sikh saint whom he used to visit regularly. During one of these visits, the discussion had turned to the current state of the world and to the seemingly inevitable catastrophes that were just down the road. Swami wondered how the devotees would be saved. "The devotees will be saved," the saint replied, "because they do not possess anything."

"Now, even husband and wife want to possess each other," Swami continued. "Originally, the relationship was free. The man experienced his wife as an embodiment of the Goddess and thus saw other women in the same way. Likewise with the woman. Through living, one became the whole. The mind was peaceful. The external relationship between man and woman mirrors inner marriage and consciousness, the *Shiva-Shakti*. It is a perfect relationship." He walked down the mountain, not upsetting anything in his path. "The saint," he continued, "is married to God, and he shares his life with the whole world. Whoever comes in his path is considered to have been sent by the Lord."

"Last night," Swami said, "I had a dream about you. You were driving a cab with a white robe on." He laughed. I had told him of my stint as a cab driver, the hold-up and all, how one night on 109th Street and Central Park West, a rather dapper-looking passenger had put a gun to my head. Swami had thought the whole incident to be rather funny, adding that I should have said when the money was stolen, "Fine, but one day you will have to come and pay it all back." For Swami, the dream indicated that one could indeed work in the Western world if one's work was truly dedicated as service. He emphasized the importance of thinking in terms of service. When I asked him what I should do about earning money, he replied that worrying about money indicated that one did not yet have "one's power."

We came to a large house in the suburbs of Dhera Dhun. The whole family came out to pay their respects to Swami. After being fed and fawned over, we wound up in a drawing room with a retired army general. With the long-term illness of his father, the general had become increasingly interested in spiritual matters. He wore tan slacks and a Western style polo shirt. Sitting up intensely in his chair, he asked Swami question after question. He wanted to know how long he would have to meditate in one sitting in order to make progress, and he gulped when Swami casually said that it takes at least four hours of continuous sitting before the mind even begins to settle down. "We are uncomfortable with meditation," Swami went on, "just as a person coming in from a lit room cannot see in the dark. Gradually, if one persists, the room and the objects in the room become visible. But one must go within! When you meditate, forget your own body and forget all other bodies."

"Remember," Swami turned towards the general, "the entire creation is taking place on the stage of death." He paused and turning towards me he spoke sternly. "To attain meditation, one must completely forget about tomorrow."

Later in the afternoon, Swami went in and spoke with the general's father. He had broken his hip and was unable to turn over from one side of the bed to the other. He had already been severely weakened before his fall and was practically unable to speak. His body was frail. The few strands of white hair lay sparse on his head, and veins bulged from his forehead. Still, he opened his eyes and clasped his hands together in *namaskaram* at the sight of Swami. Swami stayed alone with him for about an hour.

On the way back, Swami took a detour through the woods to a special Devi *mandir* where he used to spend nights doing sadhana. There are thousands of such little temples all over India. Some are nothing more than stones or little enclaves by the side of a road. This one had a simple white dome which was mounted on a concrete support in the midst of a clearing. A triangular red flag waved in the air from the dome. In front of the shrine was an opening where a small interior altar could be seen. It contained red *kum-kum* powder and other articles of worship. Swami bowed respectfully by the altar and circumambulated the dome. The *pujari*, who took care of the temple and lived in a small dwelling nearby, came over and saluted Swami. They began to speak in Hindi. The man went off and returned with a bag of sweets, prasad from Ma Durga. Swami took the bag and handed it to me. Then we all sat quietly by the mandir for awhile.

As we left, Swami explained that while temples might appear to be similar, certain ones marked areas which were definite centers of power. Such areas were filled with the force of their presiding deity, and were thus auspicious for meditation and pilgrimage.

We walked upwards through curving pathways which cut through the mountain ranges. The air hung clean. You could look down thousands of feet and see nothing but open land and sky. Swami told me how, during his twelve years at the ashram of his guru, he would periodically sneak away and go alone by Ganga side. Some of the ashram members would become upset over this practice, but his guru never said a word. When the teacher passed on, the ashramites banded together into an organization. Swami went off alone.

He lived by the Ganges for many years with nothing but a blanket and a walking stick. One day, while walking by the river, he saw a young boy seated in meditation. Drawing close, he caught sight of the boy's upturned hands. A few days later, while walking through a town, he turned towards a bookstall where a book on palmistry caught his eye. He opened it arbitrarily, and saw a page displaying the same hand markings which he had seen on the boy by the river. The book explained that this type of hand belonged to that very rare soul who would give his blood for God. The boy was Sripad-ji.

They met again, some years later, at the *Khumba Mela*, that great gathering of holy people which takes place every twelve years at Prayag by the confluence of the Ganga, Jamuna, and Saraswati rivers. Sripad asked Swami if he would like to go for a short walk and mentioned that he might want to bring his toothbrush along. Swami went through his belongings, found the brush, and left his other things stacked up against a wall. They walked out of town into the jungle and returned five months later.

They traveled together for a number of years after that. "I taught him English," said Swami, "and he taught

141

Babas at
Devi Temple

me about God." Swami knew his way through the mountains in the dark. He walked quickly, a characteristic I had seen in a number of sadhus. While walking, he told me of his visit with Nisargadatta Maharaja, a realized being who used to live and hold court in the red-light district of Bombay. As the Maharaja had no desire either to go anywhere or not go anywhere, he remained where he was, and disciples gathered around him. He asked Swami what the saints were doing in the Himalayas these days. "They remain steady," Swami had answered. Later the disciples began discussing the meaning of the creation. Nisargadatta Maharaja said it had no meaning. Swami said that the meaning of the creation, the manifest realm, was to renounce it. That really grated on me, but I remained silent.

We arrived at the first paved road. Everything was dark. The lights from Mussoorie shown down through the hills. An occasional truck would rumble by and we would go off to the side of the road. Swami stood erect and walked alone. He was a true *sannyasi*, an ascetic firm and free in the renounced order of life. Perhaps I loved him more than any person I had ever met. But even after a thousand Ganga baths and meditations emptying out into the endless Himalayan air, there would be something left undone. I loved his way, but I could no longer walk in the way of another. I would have to find my own.

The night passed peacefully. We drank tea, and I had two rusks. I mentioned to Swami that often, when I looked at him, I would see his face change into other forms, incredibly deep and noble forms. "A spiritual person," he said, "is rarely alone."

"Do you mean that one's guides are always present?" I asked.

He just nodded his head slightly.

"How can one know a true guide? Every other teacher claims to be an avatar or a world savior," I complained. "Whom should you believe? The mind can manufacture anything."

143

Swami laughed softly. "You must understand India," he said. "Here, everyone is avatar." He paused for a moment and then cited a verse from the *Bhagavad Gita*. "When your mind has passed out of the dense forest of delusion, you will forget everything that has been heard or will be heard."

The next morning before I took my leave, Swami walked around the backside of the mountain and picked some special Himalayan flowers which never wilt. "Keep these with you," he said, "Hari OOOMMM."

Calcutta

On the way to Calcutta I stopped off and saw the Taj Mahal. It was awesome, perhaps the most beautiful building in the world, and it was not a temple of God but of a woman. The traditions generally posited a choice; the holy temple of the Lord or the longings of the human heart. But this magnificent structure, pale against the sky of Agra, spoke of another way, not of choice but of completeness.

The symmetrical walls, long walkways, and gardens lead the mind into meditation, into the perfection of Allah, the Spirit Most High, whose crystalline wholeness was reflected not only in the temples of ornate architecture, but in the road and its travelers—in the shawled women with mystery eyes, in the peoples held and nourished together, in the passing dynasties, and in death.

Calcutta was filled with Gothic buildings and paved streets from the British Raj. Under the main bridge over the Ganges, huge vats of black sludge poured out from large sewer pipes, which served as supports for the tens of thousands of squatters who lived there. Here the Ganges finally empties into the Bay of Bengal as do so many of India's rivers. Here the river reaches its end, Harrapans,

Calcutta

Hunas, Hill-people, Hindus, Moslems, British, Bengalis, Tamils, Tibetans, so long the flow and the struggle, only emptying out in the end into that finality which stands so far from one's vision.

The stuffed-shirt splendor of the old empire stood like a hazy phantom background to the incessant movement of street walas, pedestrians, beggars, and businessmen. Little mini-buses in which men and women had separate sitting areas rolled through the streets. Long green coconuts were being sliced up on every corner and sold for a few *paice*, their sucked out gourds piling up by the side of the road. Bobbies in white uniforms, short pants, and long black hats gave the streets an air of an English comedy spoof.

The Bengalis, while notoriously short, held themselves in proud strength and dignity. For Bengal in northern India was the land of the intellect where, it is said, a few chosen Brahmins were sent to produce pure progeny. From their lineage came the Chatterjees, Bhattacharyas, and other impeccable high castes of Bengal. The great founders of religion and inspirers of the masses, Chaitanya Mahaprabhu and Ramakrishna Paramahansa, were both from Bengal, as were the modern poet-sages, Aurobindo and Tagore. Some people attributed the evolved Bengali mind to the coconuts which were considered to be a high-class brain food.

The train station at Howrah was filled with advertisements for various miracle cures and for legal abortions costing seventy-five rupees, less than eight dollars. One of Mother Teresa's nuns passed by, standing out from the crowd in her clean white costume. Looking at her, I remembered the hard, strained face of Mother Teresa as she ordained new members into her order in the Bronx. "Love is only real when it hurts," she said.

I looked around the station. It hurt everywhere. Not only here, for if Calcutta was comparable to any place it would be the South Bronx, rags and burnt-out rubble within view of the Empire State Building. But at home you didn't see it somehow. You were desensitized from

the world that gets hidden in some corner. And here
there were so many, and there was no space. Then, there
was the glaring dead heat. The poor and the lame stalked
the station; some without hands, others with their faces
disfigured. The nets of longing seemed to be irrevocable.
Who then could take the step backwards and dare see it
all as "pathways of souls through time," as the dream cir-
cumstance of earth experience? Who had absolutely no
fear of hunger or loss in any form?

Ramnathapur

I had hoped for the darshan of Prahlad Chandra Brah-
macari, the white-haired, cherub-faced guru who worshipped the Divine Mother, and who would hold his
index finger high in the air indicating the "One," beyond
all difference, death, and suffering. He had gone, however, to visit the village of his birth, so I met one of his
disciples, a sadhu with long, shiny black hair and a
smooth brown face, who took me into the ashram.

He spent most of the morning decorating the image of
Kali which was installed in a small outdoor temple,
painting her with *chandan* paste made from sandalwood,
offering various articles, chanting mantras, bowing before
the deity, rising, and again prostrating himself fully on
the floor. Sitting beside the altar, he wound his sacred
brahmin's thread around his finger and made various
complicated hand mudras. Next he held the thread between his thumb and index finger and touched water from
his *achman* cup to various parts of his body. The *achman*
cup holds the water used to purify one's body before con-
ducting any formal prayer or ritual.

When it was all over, he sat down on a small porch
under the shade of a spreading tropical tree. He seemed
to be friendly and open, so I approached him and asked,
"Who is Kali?"

After some hesitation, the young sadhu answered. "Only a saint can tell you." Then he proceeded to tell me. "Maha Kali has ten arms for the ten karmas," he began. "There is a Kali with four arms, and Durga has two. The people are afraid to worship her, but then they are more afraid not to worship her."

Once a year, during a full moon puja in Calcutta, herds of black goats are slaughtered to appease the wrathful, blood-drinking goddess. The meat is generally taken by those of lower castes as her prasad. Her terrifying aspect has also been exploited by roving bands of "tugs," from which the English word "thug" is derived. These thugs would capture wandering strangers and offer them as human sacrifices to the Goddess.

And yet there were others, like Ram Prasad, who had broken through the veil of terror into a free expression of love, playing with life and death, free from fears of "I" and "mine." They had fully embraced the darkness and terror. Their egos had been burnt to ashes.

This particular ashram of Kali was utterly tranquil amidst the green tropical atmosphere. Throughout the morning, villagers came to ring the temple bells and left fruits, flowers, and bags of rice. There was no urge to activity here, no idea that anything important was happening in the external world. After all, as my sadhu friend finally explained to me, "She is everything."

We walked past the ashram, past the rice paddies where husks of grain were drying on burlap bags. Humpbacked cows stood lazily munching grass in the fields. We reached the main road, and I said good-bye to the good sadhu who helped me onto a mini-bus which headed for town.

The sun was setting over Bengal. Our bus passed by the thatched roofed huts and rows of people squatting at their make-shift toilet in the fields. These villages had been created for self-sufficiency and as part of their tradition stores of grain were kept for hard times. The British had emptied these granaries. Even during the worst

147

famines, freight loads of grain were being shipped to England.

If the plugs were pulled and the paper flow stopped, if the oil flow stopped, if the wheel bearing factories shut down, there would be chaos. Entire cities would go under. But here, among the paddies, coconuts, and temples, life would continue to flow, like an underground stream.

Dakhineshwar

After visiting the birthplace of Chaitanya in Nabadwip, I traveled to the temple of Kali in Dakhineshwar, where Sri Ramakrishna had resided, exhibiting as Chaitanya had symptoms of *Mahabhava*, the highest degree of intense, ecstatic love of God.

The temple was situated within a large complex which was always filled with pilgrims. One side of this complex faced the Ganges where there were twelve Shiva temples lined up together, each housing a lingam. The main entrance faced a large ghat on the river where shops sold flower garlands for pilgrims to offer to the deity. Inside the compound was a small temple of Radha-Krishna, and rising high on a large platform was the temple of Maha-Kali, the Divine Mother, with whom Ramakrishna used to converse and to whom he had offered his life.

The Ganga, drifting slowly towards the end of its journey, was filled with dirt and slime. After bathing and being covered with its waters, you felt a lingering aura that tingled. Whether it was a disease from the gray sludge or a feeling of purification was debatable. But even as the film dried over your body, it was hard to think of anything else here except God and the God-man.

Ramakrishna's room was simple and bare. The same bed remained in the place where he used to sit and talk to disciples. On the walls were pictures of him and

Dakhineshwar

Sarada-devi, his wife and consort whom he worshipped and looked upon as the Divine Mother. There were also some photos of the major disciples who had started the Ramakrishna Mission. Many people were in the room, but there was no hurried atmosphere of tension. Respectable-looking men in white shirts and business slacks were sitting on the floor with their legs crossed in prayer and meditation. I sat down beside them. They were absolutely still, and it seemed as though they could have remained that way for hours. Into this quiet, unruffled air came a luminous feeling of otherworldliness, similar to the presence of Ananda Mayi Ma.

Here the gentleness and the terror of the Mother met, creating a strong desire for sadhana and for the life of divine love. The essence which filled the bare room was the presence of the master. Infinite forms of embodied beings—naked and smeared with ashes, in cathedrals and in caves, worshipping in Vrindavan and non-worshipping as meditative Buddhists and Yogins—all these were the Mother's children. She bestowed gifts on all according to need, filling all space like Ganga. She was the universal Mother and also the imageless, formless silence, full-blown in loving; full-blown in frightening, the erotic, expansive, many-armed Goddess, holding her weapons in frenzied movement with her dancing arms:

> All paths are thine own; embrace them
> as wind embraces tree,
> as heaven embraces earth,
> as space embraces form,
> ever reverent, loving, and free...
>
> That thy soul would hear
> the song of sound as well as silence,
> that thy heart would scale
> the peaks of form as well as emptiness,
> that thy being may rise among the worlds
> and yet remain in nothingness.

149

Tiruvannamalai

Ramana Ashram was quiet and still. Arunachala Mountain rose over the town in the darkness. Arunachala was
Shiva incarnate, and Tiruvannamalai was the town where
Ramana Maharshi had sat in the perfect stillness of
absence. I remembered his words:

> The truth transcendent, first and last
> Is the experience of pure being
> The awareness in the heart of perfect stillness
> The fact behind the fiction of the "I."

Here the illusion of activity might dissipate, as with
every passing thought came the question, "Who is the
thinker?" leading back towards the brink of original desire. *Brahma-vicara*, questioning into the self, was the
method Maharshi had used to dissolve the webs of reac-
tive thinking. And Maharshi remained, open-eyed like
mountain stillness, immovable, all penetrating.

I spent the night in the ashram and played pilgrim
during the day, visiting the holy sites of Ramanashram as
well as the Durga Mandir across the ashram road, and a
little known tomb of an old Islamic saint, Hadji, which
was tucked away in a corner of the town. In the late
afternoon, after a long trek around the mountain, I came
to the carved walls of the Arunachala-Ishwara Temple.
In the afternoon sun they appeared to be light blue, arching upwards like a glorious fortress to Lord Shiva, the
bearer of the trident. Inside the compound were a number of smaller temples to Shiva's divine children; Ganesh
and Karttikeya. Rows of Shiva lingams lined the darkened stone corridors of the inner temple. The hollow
darkness stirred the primal mind, which in its hazed
origins understood the lingam's meaning.

Priests and pilgrims hovered around the inner sanctum
where lights and garlands were offered to the huge
lingam in the center. The sounds of bells and scuffling of

Arunchala-Ishwara Temple and Compound, Tiruvannamalai

feet echoed in the dim atmosphere, the center of
Arunachala.

After leaving the temple and asking for directions a
number of times, I was led by a group of young children
to a small cottage on a side street by the temple. I was
embarrassed and didn't want to knock on the door, but
the children began yelling and calling out "Ram, Ram."

The door opened and I knew it was the yogi immediately by the turban of rags wrapped around his head. He
looked younger than in the pictures I had seen and had a
sweet, high-pitched voice. Smiling, he ushered me into
the house. "Sit right here," he said, marking the spot. I
sat. He walked over, lit a candle, and placed it right in
front of me. He then returned to his seat about ten feet
away. His "seat" was, in fact, a pile of old newspapers.
Although the room was dark and dingy, I could see newspapers scuttled almost knee deep across the entire floor. A
dog walked about the room wagging his brown tail but
making no noise. Yogi Ramsuratkumar, the hidden saint
of Tiruvannamalai, sat waving a fan with one hand. I
had sent him a telegram, informing him of my visit, but
was still dazed by my fortune of being able to find him.
Suddenly, the lights in the town blacked out, and we
were left sitting in the darkness with just one lit candle.

The Yogi lit another candle by his seat. He looked like
Lord Brahma sitting on his throne of newspapers. He
waved his fan in sure, even movements, and as the air
came in my direction, I began to feel waves of energy rising in my body. This rising sensation became extremely
intense, and my well rehearsed poise and coolness began
to disintegrate. I had wanted him to see me as a true
seeker, as a sadhu from the West who knew the ways of
India. But now I felt my entire life, with all its conflicts
and calamities, rising before my vision. With each wave
of the fan, another irritation, another desperate feeling
emerged. I tried to straighten up and keep myself together, but I couldn't hold on.

I felt particularly torn between the currents of devotional feeling and the silent expanse of deep meditation.

Tiruvannamalai

One experience seemed to cancel out the other, and I
didn't know where to turn. Each succeeding wave from
the Yogi's fan just furthered the flames of anxiety.

Then Swami abruptly stopped fanning and asked for
my name. Now I had collected many names over the
years, so I asked him which one he wanted. He asked for
all of them and wrote them down on the back of an envelope, checking for the spelling. He peered over at the
envelope in the near dark. I wondered what he was doing
but didn't dare speak. Then he began waving the fan
again, and I felt even more uncomfortable. This went on
for about ten minutes until his words broke in.
"So...what can this beggar do for you?" He always
called himself "this beggar," and he looked like one. He
wore layer upon layer of tattered rags and had a long,
white rumpled beard. But his skin was smooth, his eyes
shone, and an aura of power generated from his body. By
now I was ready to explode. Losing all composure, the
words rolled out of my mouth, "Swami, can you heal this
wound inside of me?"

The Yogi laughed. "People believe that this beggar can
heal wounds!" He laughed again in his high voice which
echoed about the room before dying into silence. Sitting
in the stillness, totally uncomfortable, I tried to recover
by straightening my back and making myself as firm as I
could. I didn't want to waste the time of this great
mahatma by making an emotional mess out of myself! I
kept looking at him surrounded by his mound of papers
as he sat calmly waving his fan.

Someone entered the room from the back, a devotee
who made obeisances to Swami. Swami spoke to him in
Tamil and he left, returning with two cups of milk and a
bowl. One cup was set before me and the other cup and
bowl before the Yogi. He signalled me to drink and then
poured some milk into the bowl for the dog, who ran
over and began to lap it up. The dog, I had heard, was
named "Sai Baba." After finishing the milk, Sai Baba
bounded over to me and plopped himself in my lap,
where he remained for the rest of my visit.

I remembered Swami's laughter and the words, "People believe..." "Actually, Swami," I said, "I don't *believe* anything." Upon hearing this he raised his hands towards the ceiling and exclaimed, "Father's Grace!" The silence returned. The city lights came back on, and I was able to get a better look at the room. It looked shell-shocked. All sorts of papers and articles were strewn about. The walls were peeling, and the Yogi had an array of paraphernalia about him; a fortress of cigarette butts, loose coins, bits of paper, teacups, fans, and magical-looking items of whose purpose I had absolutely no idea.

He then began to speak about the paths of devotion and knowledge. "Yours is *bhakti bhava*," he said, pointing out that the meaning of one of my Indian names was "To Adore." "All other sadhana can serve as a foundation of bhakti. Whatever you do, do as an offering for your *Ishta* [worshipful deity]."

He paused for a while, picked up his fan, and went on. "For a *sadhika*, a spiritual aspirant, name and form is very important." He repeated himself forcefully, "Name and form is *very* necessary for the sadhika." The Yogi sat up erect now, holding his fan like a scepter exuding rays of energy. I felt myself being lifted, becoming light and airy. Then he spoke again, but in a different tone of voice, a voice filled with wonder and whimsy. "But...for a *siddha* [a perfected being], for a *siddha*," he repeated for emphasis, "Name and form is...." And then he burst out laughing, howling with a crazy elation that shook the entire room, exploding on and on, exploding my mind, exploding the room, exploding everything!

A while later he continued in soft tones, "Go beyond name and form. For a siddha there is no conflict." No conflict—I felt a tremendous power in those words. I returned to silence and now rested there more easily.

Swami then began to sing in a thick loud voice, "Jaya Hanuman, Jaya Hanuman, Jaya Hanuman, Jaya Hanuman." He sang for a few minutes. "There, is the wound healed?" He smiled. We sat for a while longer. He suddenly leaned over, looked at me directly, and said, "So

you don't believe in anything? You don't believe in
Raman Das?" I almost fell down. Raman Das was
another of my Indian names, short for Radhika Ramana
Das, which was an epithet for Krishna as the enjoyer of
the bliss of Radha. He did it, he got me. How could I
not believe in myself whose name was a name of God? I
was unmasked. It was all over.

Later Swami repeated his first question, "What can this
beggar do for you?" There was nothing more to ask, but
I wanted to remain in his presence longer so I thought of
something to say. I told him how concerned people in my
country were over the volatile affairs of the world and
how various groups were preparing themselves for a holocaust. I asked him what he thought about it all.

"I too am concerned," the Yogi replied. Then he
paused, one of those long pauses where you wait on edge,
until his face broke into a gentle beam. "Father loves humanity, and all He does is for good. Father loves humanity and Father will save humanity." He paused again, "If
Father has humanity blown up, that too is Father's
Grace! If He wants to save humanity or destroy, that is
good. Everything is Father's Grace. Just remain at the
feet of the Father."

Sai Baba in Bangalore

Before leaving Tiruvannamalai, I returned to the residence of Yogi Ramsuratkumar and called out, "Ram,
Ram." He came out smiling, took both of my hands, and
held them together in his for a long time. I looked up at
him as he stood on the porch, and saw nothing coming
through his eyes but beams of love, pure acceptance.
"May Father grant you all success on your pilgrimage,"
he softly purred as he rubbed my head with his palms.

He had polished me off like an apple, and I arrived in
Bangalore filled with expectations at seeing Sai Baba.
Along with my rising expectations came the acute desire

for something to happen, for some miracle or revelation to manifest. I had fantasies of Sai Baba walking over and picking me out of the crowd, embracing me as his long lost son, looking deeply into my eyes and unveiling my destiny.

When I arrived at the mushrooming ashram complex, it was raining heavily, but even in this rain crowds poured in by the thousands seeking the darshan of Sai Baba. His miracles and superhuman powers of manifesting materials from thin air had made him renowned throughout the world. Perhaps even more impressive were the schools and colleges inspired by Baba which had literally risen from the dust of South India over the past few years.

I walked around the ashram which was also called Brindavan, catching bits of conversation, reading pamphlets, and observing the swarm of activity. It seemed quite possible that Sai Baba might indeed resurrect the Indian nation and guide it back towards the *Sanatana Dharma*, the eternal religion of the soul.

As with any powerful figure, there were storms of controversy surrounding Sai Baba and his mission. There were negative insinuations and suspicions about his personal life. There were attacks upon his organization and its methods, and there were all kinds of cross arguments over his claim to be the Avatar, the incarnation of God descended into human form for the upliftment of society.

The decriers claimed that the materialization of objects was a fraud, or that Sai Baba had sold himself to the djinns, spirits who created the miraculous manifestations. Others declared that he was committing the most heinous sin of impersonating the Lord. Then, however, there were the countless numbers whose lives had been transformed by Baba's presence, who had received the gift of devotion, and who radiated the love energy of God. "In India everyone is an avatar." I decided to remain open; to open my mind and heart, and receive whatever was to be received.

There was nothing but rain all day. Upon hearing that

Sai Baba in Bangalore

Baba would not come out, the crowds began to disperse reluctantly. Behind the shops along the main road were small dwellings shared by pilgrims. For a few rupees, you were given a space on the floor and the use of a toilet around the back. We all huddled together in a small room as it rained. I found myself sitting next to a perfectly ordinary Indian gentleman, middle-aged, and wearing Western clothing, or to be more accurate, the Western costume adopted by the emerging Indian middle class. He could sit on the floor with his legs crossed much more easily than I could. As I was trying to adjust my seating position, I heard him softly musing out loud. "I did not believe in Baba until Baba came to me. I was doing my puja and he appeared right over the altar. I said, 'Baba is this really you?' He told me to come to Puthaparti, so I came to see Baba. I waited for days and Baba didn't even look at me. Then Baba saw me and gave me a blessing. . ." The man went on speaking.

After the rain, as I was carrying my things over to the guest house where I had managed to finagle a room, a woman with light gray hair and wearing a blue sari offered to help me. She was an artist, and told me how one day while sitting in her living room in Australia she found herself sketching an image which fascinated her; a large face with a bushy Afro and wide glowing eyes. It was not until sometime later that she found out that it was the exact face of Baba. I soon found that everyone had a Baba story around here. "Don't think you are here just by chance or by your own will," I was told at least a half dozen times. "You only come here if Baba sends for you."

The air had cleared by morning. The sun came up early and dried out the grounds. It was bhajan day at Brindavan, and thousands of devotees and visitors were cramming into the area to participate in the devotional chanting which would continue uninterrupted throughout the day. As the crowds grew and the instruments were brought out, the *seva dals* or ashram workers (who were, in fact, the ashram police) dressed in white, directed the streams of humanity into their proper places, men on one side and women on the other. The rows of seated pilgrims formed behind chalk-drawn white lines which surrounded a sitting area by a large green canopy. In the center, under the canopy on a raised platform, was a smooth, white statue of Krishna curved in his three-fold bending form, holding his flute. Now Krishna was supposed to be dark blue and Baba was brown black, but this was India and there was room for everything.

Pilgrims walked up, one after another, and offered fluffed yellow flower garlands to the deity. The lines expanded outward until the whole area was filled. Tension mounted everywhere as people jockeyed for position on the lines. Three women dressed in bright printed saris came walking up the center aisle. They carried themselves with that conspicuous air that told you they were members of Baba's "in group." They dressed the deity in a cool green dhoti and wrapped his top in a pink silken cloth with a gold border. Then a red carpet was rolled out, and Baba's chair, along with his paraphernalia, was carefully placed at the end, down in front of the platform.

The bhajans began, and the whole crowd swayed to the chanting, "Manasa bhajane guru caranam..." These people all knew hundreds of bhajans, and with each new one the intensity and expectation mounted. Then, the first sight of Baba, as his deep red dress and Afro hairdo emerged from the archway at the end of the compound. All heads turned towards him as the singing soared in volume. The air was like heavenly rainbows, Vishnuloka, the divine planet of the Lord. The ground appeared as touchstone. Everything was transformed. Baba walked out, slow, serene, imperturbable, picking up the border of his robe with his hand. And as he walked up the red carpet, there was an outpouring of love. It was a love that you could really feel, a love which even extended past the awesomeness of such fullness and power being present in one person.

Baba sat in his chair, moving his head to the rhythm of

the chanting. Every once in a while he would slowly survey the crowd. For a moment his eye passed over me, or so I thought, and I felt a surge of tremendous power lift me upwards through my body. Rising from his chair, Baba slowly walked around the pathways in the arena, stopping in front of a few lucky souls, taking the letters which they had prepared and were waiting to hand him, blessing others, and occasionally stopping to talk to a person. Then he would turn his hands around and around, palms down, and manifest something out of the air, handing it to his devotee.

After seeing him in person, no matter how many movies you may have seen or books you may have read, you are awe-inspired. For the rest of your stay in Brindavan it is impossible to think of anything else except Sai Baba. You begin to wait for the next darshan, and time melts from one darshan line to the next. Soon, it becomes apparent that the main sadhana here is waiting. You wait two or three hours in the morning and evening for Baba to come out. When he appears, you wait, hoping that he will walk over to where you are sitting. When he walks over to your side, you wait, anxiously hoping that he will stop in front of you and acknowledge your existence in one form or another. Then, when he walks right by as if you don't exist, you wait all night figuring out what you did wrong and scheming to get a better place on the line the next morning.

Some of the regulars had devised elaborate systems to communicate with Baba. If he put his left foot forward and turned towards you, it meant one thing. If he walked to the right side during the last two darshans, it meant that he would stop left of center in the next one. It was clear, however, that Baba was in complete control of this act.

I began to feel myself getting involved with the energy of the place. I would walk back to my room and nurse my mind, wondering why Baba had passed right over me again. He had not even given me so much as a glance. I wanted to see Baba, and I wanted him to see me. "But he hears and sees everything," they all said. But he did not see me. I had absolutely no concern for anyone else at the place, for why they had come, for their hopes or seeming predicaments. Seeing the grotesque intensity of my own self preoccupation, I began a ritual of self-condemnation and took vows to become holy.

Still, I could not stop thinking about Baba, being aware of his presence or omnipresence, whatever it was. In the morning I asked Baba in my heart to take charge of my consciousness while I was here, and to help me out of the confusion arising from desires for personal gain and happiness. That afternoon, while resting, I dreamt of two opposing street gangs coming together to celebrate a marriage between members of the rival communities. I awoke with a sense of consolidation.

The next days were spent waiting; either indoors, waiting for the rain to stop, or outdoors, waiting for Baba to appear. I was completely alone and turned inward. I had made no friends here and wasn't looking for any. I wondered as I waited who on earth these gurus were anyway and which one was my very own, my Sat Guru as they say. Was it Sai Baba, or Sripad, or Hilda, or Jnanananda? These questions bubbled up and burst open as I waited for an answer which my heart knew did not exist.

Every morning I took my place and waited for hours. The rain was always a factor. If it came, Baba didn't. I sat and gazed off at the deity of Krishna, chanting his name, but that became tiring. If you wait long enough, your emotions just wear out. All I saw was a statue. In fact, the whole ashram began to look like a circus with each of the various Gods having his own side show.

I slumped down, then walked over to the bookstall and out of impulse picked up a book containing small aphorisms of Baba's. Tension had begun to increase on the lines. Anticipation came in waves. He was coming again! At that precise moment, I don't know why, I picked up the little book and opened it at random. As Baba appeared through the archway and began moving towards our area, I scanned the page and found the following:

Sai Baba in Bangalore

> It is really a source of amazement, this creation and
> the wonder with which it is filled. But considering pre-
> sent conditions, there are very few who watch for the
> Light, and who are guided by the Light. So, instead of
> following this person and that, and taking devious
> roads and getting lost, it is best to place full faith in
> the Lord Himself, and rely on Him as the only Mother,
> Father, Guru, and Guide.

What a power came through these words, a grace that
meant more to me than Baba emerging in all his glory.
Baba stopped and surveyed the crowd, wiped his
forehead with a handkerchief, and slowly moved on. The
crowds were leaping at him, and he soothed them all
with his raised hand. I noticed a young boy seated not far
away from me. I angled my eyes over in his direction to
catch the last words of a letter he was writing to Baba.
He was signing it "Your loving devotee." I felt a wonder
at this young boy's devotion, appreciating the difference
between his full faith and my own curiosity. Baba took
his letter.

With each darshan, I would notice Baba's face and
glazed eyes which were seemingly so removed from the
human plane. He had an ever-present aura around him
like a rainbow. He never lost his presence, his equanimi-
ty, even in the face of thousands. And the more I looked
at him, the more I had to look inside myself, had to ex-
amine my own motivations for being there, had to witness the chaotic desire-centered contents of my own mind.

So the days were spent waiting, and in the evenings I
would be brought back alone to my room, where the rising fragrance of incense seemed to carry his presence like
vapor.

To accept all as the grace of God, even rejection, I
knew that this alone would free my mind and heart. And
there was a slow, vaporous feeling of an emerging presence, a presence which could lift all complications, a
presence which could never be achieved through effort.
Beyond my cocoon of desire and ambition, which I now
saw as the major source of my pilgrimage, all experience

seemed to grind me down and down to this; to have
nothing, to want nothing, and to be no one. The entire
oscillation, what to do and what not to do, could vanish
like a shadow. My heart was becoming sober. It saw that
the rush towards elevation placed the single greatest
burden upon true understanding.

More days went by. Baba's presence was all-encompass-
ing. And with each day, I felt my resistance ebb. Now
there was nothing to cling to. The presence came again
that night. In a dream a figure came and touched my
forehead. I felt a cracking sensation, like the breaking of
a web, and went flying out of my forehead. I let off a
silent cry. I was terrified. I didn't know what was happening to me. Everything was dissolving. But I was able
to finally hold with it, even relax into it, and soon felt
myself to be kind of hovering around, floating free in
space before slowly sinking into my body again. The body
was still sweating as I awoke to the words "Beyond name
and form."

Yet on my last day at Baba's I was still waiting. Baba
appeared, and came towards the side where I was sitting.
There were a few of us lined up in a row with packets of
vibhuti, holy ash. We held them up for Baba to bless as
he walked by. He lightly touched the first person's
packet, the second person's, and hovering just above my
own, he passed right over me and proceeded to bless the
next one. His face, during all of this, was totally serene,
impersonal, and filled with bliss. I gazed at his form as
he walked away.

Walking with my bags towards the central check point,
I thought of Yogi Ramsuratkumar's chiding me. "What,
you don't believe in Radhika Ramana?" You don't believe
in yourself? I paused by the window to return my key
and as I was about to exit I ran into a message from
Baba. It was written on a blackboard in chalk. They
were waiting to hang it up outside the bookstall as the
lesson of the day;

> Do not look for God in the world around you. Have
> firm faith that you are the atman [the eternal soul],

that the divine spark is in you. Move out into the
world like heroes whom success does not spoil and defeat does not discourage. There is no need to call on
Him to come from somewhere outside of you. Become
aware of Him as your inner self.

Bombay—The Return

The plane to Cairo was not leaving for a few days, so I
stopped by the Elephanta Caves and saw the massive
carvings of Lord Shiva and the great stone lingams of
power. The relief carvings had been damaged by the Por-
tuguese who had used them for target practice. Nevertheless, the imposing figures cut right from the caves
themselves gave off a silent power and a feeling of
hallowed permanence within their natural sanctuaries.

The next day, I took a bus to Ganesh Puri to visit the
Muktananda Ashram and to pay my respects at the
samadhi of Bhagavan Nityananda, the great *avadhut*.
The ashram was filled with statues and fountains. There
was one statue of an elephant being saved by Lord
Vishnu. This portrayed an old story from the *Puranas* in
which Lord Vishnu flew down on his bird carrier Garuda
in order to rescue an elephant who was being pulled into
the water by an alligator. The elephant, the head of his
herd, had been deserted by his family and friends as the
alligator slowly pulled him down. In a previous life as a
human being, the elephant had been a great devotee of
Vishnu, and at the last minute he remembered this and
called out the Lord's name. Naturally the Lord showed
up like the cavalry and severed the head of the alligator
with his razor-sharp Sudarshan disc.

I met a Western sannyasi who had received his initiation into the renounced order of life from Baba
Muktananda. "The minute I saw Baba," he said, "I knew
he was my guru. I had been wandering around India for
eight years. If I didn't like a place, I would just pick up
and leave. But I knew somewhere that I would have to

stop and say, 'This is it.' He asked me if I was a sadhu, and I said I was. So he asked me if I would like to take initiation. He gave me *sannyasa*, and I gave him my life."

We went down the road to the samadhi of Bhagavan Nityananda, the most revered guru of gurus. It is said that he appeared under a tree and kept rising out of his body. It didn't look as though he'd be on the planet too long, so they force-fed him some crow's meat, and that brought him down into the world.

They were playing bhajans in the temple over a loud-speaker. There was an imposing statue of Nityananda with his shaved head and tank-like body, sitting in the lotus position. Pilgrims came by making offerings of flowers and money to the altar. The temple itself passed out sanctified food; sweet halvah, warmed and served on small leaf plates. After sitting in the temple, we went back to the natural hot springs which were collected in three bathing tanks that ranged from warm, to hot, to nearly scalding. I lingered in the baths, letting weeks of ingrained dust wash off my weary body.

The shops were almost closed down. I was alone now, walking back to the ashram. At the end of the street one general store remained open. Its radio was on high and it would blast all night through the streets of the holy city, even though electricity still felt incongruous out here. At first I didn't believe it. I thought that it couldn't be, not here. The song must be coming from somewhere else, a memory perhaps.

As I came to the store and listened closely, I knew. Right here, in Ganesh Puri, as the village went off to sleep, as the Great Mother India closed her lids, as the cows lay along the road, and the chay shops put up their stalls, as the temple curtains closed, and the high-caste Brahmins cuddled in under their mosquito nets, as the hot springs rippled and the dust from the plains blew through the air, as the temples all stood in silence, and the rickshaws, bicycles, cars, and oxen lay senseless, the song came through the airwaves, through the ethers, announcing the permanence of the Western invasion, or perhaps the ability of the Great Mother to accommodate all. I stood in the night and listened to the song, "She's Got Bette Davis Eyes."

4
Egypt and Israel

Moving back again, circling towards the center, towards
Africa and the origins, where did it all fit together? Nations, cultures, and peoples, where would the world culture be born? I felt a new sense of anticipation as the
plane flew on to Cairo. Perhaps it began while I was
looking over someone's shoulder on a bus and reading the
newspaper headlines, "Egypt after Sadat." The anticipation had to do with re-entry into the world of history, the
world to which, for better or worse, I was bound. India
felt like a panoramic dream, remaining within the vast
dreaming recesses of consciousness. What would happen
to that dream, to the great myths and gods of the past, as
the rising world of technology firmly asserted itself over
the subcontinent?

But the historical world was but another mythic form,
one agreed upon by the majority of current processors of
information. What would happen to its roots? Would the
Holy Land remain holy, or would the ancient be obliterated? Anticipation. I didn't know what to expect. I didn't
know how I would feel at this coming meeting. For these
cultures still peered through our foundations, the pyramid
rising on the dollar bill, the new messianic stock in the
state of Israel popularized by the born-again Christians,
and most disturbing of all, the ever-present threat of vio-

166

lence, of disruption, of annihilation itself which hung as
if fated in the air over the Middle East.

The Great Pyramid

Egypt was very busy trying to hold its government together. The new President had his picture plastered
everywhere alongside that of the slain Anwar Sadat. In
Cairo, soldiers stood behind sandbags, holding bayonets,
their eyes expressionless before the shifting crowds.

The women with their covered heads, the cloth merchants, the drivers, and dwellers in the old city—they
seemed oblivious to it all. Newscasts and political arrangements were far off in another dimension. The dirt
roads with their squalid dwellings and grinding poor were
kept well away from the Sheratons and Hiltons. In Cairo,
if one cared to look, one would find two distinct Egypts.

And there was yet another Egypt, one whose presence
was foreign and whose memory was estranged, as seemingly absurd as the electric light shows which played
upon the pyramids. This was the Egypt that I was seeking, and I wasn't the only one. After finding a small
room through one of the innumerable street hustlers who
tried to sell me some of his uncle's "oriental perfumes," I
found myself roaming through the old city, visiting the
small Coptic churches which still stood behind unpaved
streets with their animals, noises, and rubble. I was accompanied by a young German traveler named Andreus
who kept steering me out of the churches and back into
the streets. "You've got to see God in the garbage," he
said.

The next day, in the late afternoon, I took a crowded
bus out of Cairo and proceeded towards Giza. Beyond
the hotels, light shows, ragged streets, and hawking camel
drivers, the pyramids loomed over the desert, foreboding
and magnificent. The sun was still high, and the broad-

167

*Pyramids
at Giza*

ening towers of blank stone cast long shadows over the sand. Their presence immediately captured the mind.

This was not my first visit to the pyramids. Once, on a connecting flight, instead of staying in the airport hotel for the night, I had taken a taxi out to the pyramids and slept outside on a sand dune. The camel drivers had come over to me and offered to take me up the wall for a fee. When I said no thanks, they offered to sell me some hashish. They still wouldn't take no for an answer and hung around for most of the night watching me meditate and try to conjure up visions of Atlantis—another one of those weird foreigners.

My healing friends back home had been to the pyramids and had spoken of etheric seals that one had to break through before entering the King's Chamber, as well as the specific meaning of each wall, north, south, east, and west. I readily believed all of it, but there was something more mysterious, something more foreboding than all the speculations of ancient glory. It was this feeling that kept pulling me now.

Stone riddles in the desert, geometric mysteries of an unknown age, impenetrable on the flat desert expanse. I had lain on my back all night seeing glimmers and flashes, Atlantean configurations and a strong sense of water, but the gates were closed. The time was not now. It lay under the expanse, still buried in the akashic sands.

I now entered the Great Pyramid of Cheops and slowly inched my way up the narrow passageway which led to the King's Chamber. Edging up sideways, I felt a strong sense of presence, of guardians, of mysteries which remained beyond me. I finally arrived at the great hollow room of thick grey stone. An open sarcophagus was resting by the western wall at the end of the chamber. There were a few tourists left including some space-eyed young people who were chanting and listening to the echoes reverberate and return as intensified sound. The air felt sparse and difficult to breathe. I walked over to the far corner of the chamber, assumed a sitting position, and closed my eyes in meditation. The energy in the room aligned one immediately.

Awareness expanded into a continuous presence. I was only dimly conscious of the movement around me. I did hear the announcements that the pyramid was closing and that everyone had to leave, but all my energies were gravitating towards this center, this spot where I was sitting, and I just remained still. In the back of my mind somewhere, I assumed that at the right time I would come down from this state and leave. But back further still, as the center kept deepening, I knew that I wasn't going to leave at all.

I'm still not quite sure how it all happened. I just sat absolutely still when the man came back telling everyone to leave as they were closing for the night. At one point, I got up and went into a little side cave which burrowed upwards. The doors were closed now and everything was still.

I returned to the King's Chamber and resumed my sitting position. Silence was interrupted by an abrupt noise. I was jolted and opened my eyes, but I couldn't see a thing. The lights had all gone out. The pyramid had been closed. It was so dark, I couldn't see an inch in front of me. I waved my hands in front of my face, in the fear that I had gone blind. I could not see them at all. There was no light. I rubbed my eyes with my hands. Seeing the flashes of light inside, I felt my fear somewhat lessen.

I knew that I was in here for the night so I slowly backed over to where I thought the wall was. While moving, I tripped over something and bent down to feel it. It was a knapsack. I found the strings and opened the flaps, pulling out a blanket and some assorted clothes. I felt around the side pockets. There was a book of matches. I lit one and was greatly relieved that I still had my sight. I looked around. It was the same King's Chamber, but there were only two matches left. As the flame went out, I decided to save the others.

The air was stuffy. I tried to take a few deep breaths. With those breaths came a shooting pang of fear. My mind spun off into all kinds of crosscurrents. I had set this whole thing up. I had wanted it to happen. But now there was no backing down, and I wondered if I hadn't made a great mistake. The shadows of fear drifted through my body.

With my back to the wall, I straightened into the posture of meditation. I was conscious of a total aloneness. Not a single sound could come through the thick walls. There was no light. It was impossible to see anything. As I sat, I felt that the air was becoming increasingly difficult to breathe. Perhaps the shutting of the doors to the passageway meant that no more air could come into the King's Chamber? This was my meditation.

In this dark aloneness the sinking fear continued to creep over me. It began with the lack of air but went on into other depths. The fear began to take hold of my body as the stomach area tightened. And this fear was not something new. I saw it as being very, very old. It was so old, so well covered and locked away, that I had almost forgotten its existence until that moment.

First came the fear of death. As it surfaced and began to take possession of my thoughts, I became convinced that I would not walk out of this pyramid alive. Perhaps I would suffocate in there. Perhaps some entity from within the pyramid would just walk into the chamber and do me in. Who would know? Who knew I was there? What was unfinished? What about my family and my friends? While I was going through a whole list of such thoughts, a second wave of fear began passing through me, even deeper and more vivid. Pure fear contracted my entire being, fear that was at the core of everything. There was a total loss of reference, like being blown wide open out on some astral plane between worlds, and the slightest move would tear me to shreds.

I had not glimpsed this fear in quite some time. Perhaps during a drug experience over a decade ago there had been a brief encounter with it, but now I understood that it had always been with me. I suddenly had visions of losing myself, of crossing over the irretrievable line, that thin line between madness and liberation. Floating through a maze of fears, I remembered Vrindavan where a twenty-five-year-old man was dragged into a psychiatric ward after spending the night in the forbidden grove of Niti Bandh. He had been the most recent in a long string of insanities and deaths occurring to those who dared to spend the night in there. The young man died a few days later. I saw myself, all twisted and contorted, incarcerated in some state of madness, hung up in some psychiatric hospital, being stared at by bearded white-coated men with glasses. I felt demons coming out from the walls and laughing with contorted faces, each one bringing on a new rising tremor of panic.

I was trapped, locked in. I couldn't lose control. I would be eaten alive, dissected, tortured. I went to light another match. I could still see. I tried to recover myself and looked around the ominous chamber. There was one match left. If I tried to get out of this place with it, the light wouldn't even last as far as the passageway. This was it. I had to face it. This entire pilgrimage, my whole life, had led me to this room. Very deep and very far down, I could feel the fear. It was the core, the very core of my life. I tried to align myself, to intensify my meditation and not give in.

In this aloneness, perspectives were reversed. I began to feel as if I had died and had left my body with all its familiarities behind and was floating through this dark space. The span of a single lifetime was seen as no more significant than a grain of sand on a beach, an entire life, nothing but a glimmer. And now I was back by the borders of hell. I knew that I had been here before and was now desperately afraid of being stuck here.

My mind raced through the universe, seeking a point of reference, something I had done during my life that could anchor me, give me some solid sense of reality. I felt myself tightening further. All my prayers, meditations, and repetitions of God's name weren't worth a damn in here. They only seemed to make it worse. I searched through all the seemingly good and meaningful things I had done, all sorts of experiences and their conclusions, but they all vanished. Like dreams, they were only shadows of reality. The breath was getting heavy, and I breathed deeply, struggling for more air. I tried to fight it. I would refuse to go insane. But the fear kept rising and taking greater possession of me. Sweat poured from my body, and in my mind I kept hearing the words, "Hell, hell, hell, hell, hell."

I knew I couldn't last. I would have to let go. Deep in the center, I felt something break, like the snapping of a branch of a tree. I was dying. My hands were slipping into the sea as Atlantis burned and there was no one, no helper and no God. The branch was broken. The grip was gone, abandoned into an endless free fall.

From an inner pool there arose a vaporous tone, a tone of peace. It was not the peace of the altar, nor the love of the holy pilgrim's way, but something else. It was that which had been given from the heart, that which had been shared with others during life. This vaporous feeling began to rise from an opening of great depth, and as it did, fear began to dissipate. It all seemed to be happening by itself. As the mist rose through me, I felt that if I ever resurfaced in the world of form, my only concern would be this sharing, this love, the only thing of any significance in my life.

In this expanse fear receded. I could feel it evaporating, and I felt aligned and free, floating in a silence, an absolute dark silence. Death relieved all burdens. I remained adrift, afloat in the deep silence and darkness. It was a clean darkness with no need for the light of heaven. The silence was so deep that it seemed to have its own sound. There was no more resistance, and I allowed myself to sink and sink.

I don't remember too much of anything else. At one point there was a sensation of my body being reconstructed on the etheric level, as if the atoms were broken apart and were now being put back together. Then there was another impression, like that of a radio call into outer space coming from earth. The call was a tug of subtle sexuality. I could feel it within my genitals as if someone had fastened a cord to my inner body and was pulling me down. I understood this to be the pull of the earth, a pull still programmed into my atoms. The pulling felt like a deja vu. It seemed that I was being forced to reincarnate again, to experience and work through the earth again, to grow into another form to learn again. I felt myself gradually ascending from the depths, becoming more and more conscious of the cold stone, of the rarefied air, of the dark room, of the knapsack and the blanket. I groped around and lit the final match. The chamber was still there, empty and silent. I fell asleep on the floor.

I heard a sound and the lights came on. I could not believe it. I had lost all sense of time. I got up and walked around to collect myself. I put the articles back in the knapsack and, following the light, made my way down through the passageway. The entrance to the pyramid was open, and light was streaming through. It was daylight.

An Egyptian guard came through the door and looked at me as though I was a ghost. Then he began to say, "Police, police."

"No, no," I said, "no police." He looked at me again as I was about to walk past him. He held out his hand and said, "Tip, tip." I knew I was back in the world. I reached into my pocket and gave him a few crumpled pound notes and walked out into the morning desert without looking back. The air outside was clean and new. The sun had come up over the sand. I walked past the Sphinx feeling as if I had resurfaced from the dead, but I was too tired to think about it all. I only wanted to return to my little rented room in Cairo and go to sleep. I walked past the camel drivers and merchants and then through a long vegetable market place until I found the main road and caught a bus back to town.

Entering the Holy Land

The border at el-Arish was closed. Normally I would have taken a single room for the night, but the "voice"

said to take a larger one, that someone would come and
share the room. There seemed to be no tension in the
town. I watched the movements of pedestrians and merchants, and the weathered faces of old nomads with their
camels. Bowls of steaming couscous were being sold in
the back streets at dusk by robed men behind wooden
pushcarts. Chanting could be heard coming from the
Mosques at the edge of the village, and by the sea the
moon shone through a cloudy haze.

In the hotel, men sitting around in Moslem robes
drained black tea and watched American television reruns
which had both Hebrew and Arabic subtitles. I looked
Taround the room and wondered what everyone was
fighting about.

Down in the restaurant, I ordered twelve greasy
falafels and some *tahini* for fifty cents. Some men sat in
the back around a *hooka* or water pipe, others milled
about in the streets. What did people need anyway? Some
food, a place to live, and a few friends. I loved it all.
The tea, the tablecloth, the grease, the people in the
streets, the Arabian music, the sea, and the air—everything was doing what it was doing. Drinking tea and
eating falafels was absolute activity.

A middle-aged man with glasses and a loosened tie,
looking a little like Menachem Begin, came into the dining room. He spoke to the waiters in seemingly familiar
terms and then walked over to my table. "Are you Rick?"
he asked.

"Sometimes," I answered. "And you?" He told me his
name, a name which I could not pronounce, and explained that he had booked the other bed in my room. It
had been the last one available. He had come from Alex-
andria and was on his way to Tel Aviv in the morning.
He had sped through the desert all day hoping to reach
the border but had some car trouble on the road. He
spoke about his business selling textiles, and about the
problems of his new American car. Looking around the
restaurant, he informed me that things had changed here
since the war when he was an army captain in the Sinai.

175

Egypt and Israel

From that moment, I started calling him Captain and then "Cap" for short. Eventually, he stopped his monologue, looked at me, and asked where I was coming from. I told him that I had come from India.

"India! What's a nice Jewish boy like you doing in India?"

"I really don't know," I answered. "I've been trying to figure it out for the last ten years or so." We went on talking, or rather, he went on talking as I drained another cup of tea and mopped up the rest of the tahini with some pita bread. He asked me where I was going in Israel and how. I told him that I was planning to take a bus to Netanya where I had the address of my friend's sister. He offered to give me a lift into Tel Aviv the next morning.

We arrived at the border early in the morning. The Egyptian guards fumbled around with our passports in the same way they fumbled around with their rifles. They didn't seem to take anything too seriously except the five-pound tax you had to pay to leave the country. "Imagine that," said the Captain, "they have the nerve to make you pay to leave."

Things on the Israeli side were much more difficult, with innumerable forms to fill out and checkpoints to cross. "They are afraid of terrorists who come in disguised like you," Cap explained. "I'll talk to them." The Captain walked over to a woman wearing a brown khaki uniform with a machine gun slung over her shoulder and informed her that it was all right for me to go through. "He's Jewish, I can vouch for it. He's going to visit his aunt in Netanya." We were out in five minutes, signed and sealed, and were soon speeding through the Sinai.

The desert was empty and barren. Amidst sands upon windblown sands, a rusted bombed-out tank would appear every once in a while by some brown desert brush, a reminder of what men had done to this land. Nevertheless, there was a very real feeling of living history in the air. As we crossed the border and entered Israeli territory, I was surprised to find tears welling up in my eyes and a deep throbbing in my heart. I felt as though I was coming home after a long journey, home to that site which is the conclusion of history.

Cap talked constantly as we drove on. "The tour I am giving you," he said, "you could not get for fifty, not even a hundred dollars!" Of course, he was right. He stopped the car in all the towns and settlements along the way, telling me their histories. He kept speaking of the "war" as well, that is the Yom Kippur War and the part he played in it. "I have fought in every war since 1948," he said. "When I came here from Europe, I kept my last name. Everyone here was changing to a Hebrew name, but mine was the last surviving name of our family. Everyone else was killed in the holocaust. But I gave my daughters Hebrew names. . . . No, I don't believe in God. . . . how can I believe in all that. . . but," he paused, "I believe in Israel. And if it wasn't for God there would be no Israel. But let me tell you something Rick, just one person, any one person like myself, living and working in the state of Israel is worth more than ten of your Rabbis in Brooklyn praying all day and night."

As he spoke, we passed a settlement with rows of trees and farmland all around. "Look!" Cap exclaimed pointing to the plowed fields. "We have made the desert bloom! We have made the desert bloom! So what do they want from us? These Arabs, they are our people, Semitic people. We can help them. We can build hospitals, teach irrigation, so much more than the Russians. Look, did you see those boys working in the restaurant? I know them from Tel Aviv. They told me that they wished they were back there. You see, they cannot get a good job. We have nothing against them. We are Semitic people, the same people. Go and speak to them, the Arabs, the ordinary people. They do not hate us. Then who made all the wars?" He turned and looked at me with the gleaming eyes of the true believer. "I'll tell you, it's the politicians. Do you think that the people want to see thousands of their young men slaughtered?"

Cap rambled on with the car. "During the war when

177

we crossed the Suez we had twenty thousand Egyptian troops surrounded in the desert. Their food and water supply was completely cut off. It was just a matter of time and they would all be finished. So Kissinger came in. He said that this time if we let them go, maybe they would make peace. Tell me, have you ever seen or heard of such a thing, to let the enemy who has attacked your own land go free? Who else but the Jews would give back land that has been won with their very blood! We let them go, and we found oil on that land, and still gave it all up for peace."

Our drive through the desert continued. "What is your last name?" the Captain asked. As I spelled it out for him, he proceeded to delineate my entire family lineage, telling me what country my forebears came from and what professions they worked at. "You see," he explained, "these were not the real names of the people of the Diaspora. They spoke Yiddish so that no one else could understand. The Germans, Poles, or whoever was ruling gave names according to professions in order to tax the Jews; Silverman, Fishman, Tailor, and so it goes. Now that you have come to Israel, you will find your real name."

During the rest of the ride, I was given the entire history of the Jewish race and an assortment of other historic truths. Cap let me off at a bus stop. Seeing that I had not changed any money, he gave me some shekles for the bus and a little extra, refusing to take American currency in return. He shook my hand and said good-bye.

In Netanya I told my friend's sister Gloria about my ride. She asked if he had given me the "whole spiel." "Absolutely," I replied.

"Well, be prepared," she warned me. "You're going to hear it again and again." I stayed at the house for a few days, going to Tel Aviv in the afternoons. Tel Aviv was a lot like Brooklyn College with one noticeable difference. Everyone here had machine guns slung about their shoulders. If Brooklyn College kids even saw a machine gun they would have heart failure. Here, everybody was in the army.

Gloria told me how during the Yom Kippur War, the men were coming out of the synagogue when they heard their code names being called on the radio. You weren't supposed to drive on the holy day, but they had to speed over to their defense positions since the attack had been a total surprise. It is said that as the soldiers raced to their posts there were more deaths due to traffic accidents than due to the war on that first day. "We lost five thousand boys," she said. "Do you know what that is in a country which is not even the size of the state of Delaware?"

Gloria showed me the air raid shelter in the basement. All houses were equipped with them by law. In the evening, the children came home and watched American television. I noticed a picture of a young, blond, curly-haired man on the living room wall. Next to his picture was a pair of dog-tags and a machine gun, also mounted on the wall. "My husband," she said. "He was shot down over the Suez. They sent me his rifle." The elder son would soon be ready to enlist. All citizens were required to do so, and there was the feeling of being ready to fight unto the end.

In Haifa I met Michelle, an old friend from the farm in France. She was now living in a small community not far from Mount Carmel. We had been close at one time, but Michelle had become holy and moved to the Holy Land. Since then, despite our attraction, there was always a feeling of uneasiness between us. We kept in touch, but always brought along some barrier to avoid a confrontation. That barrier was usually God.

This visit was no exception. I was invited into a small house turned "ashram," and given a place to stay with the "brothers." I was treated kindly but kept at a distance. Still, a light shone in Michelle's eyes and she always looked holy. She had been in the country for three

years now and gave me a list of holy places to visit in
Jerusalem. A few times, we began speaking seriously, but
it always ended in frustration. Sometimes I would take
hold of her hand and look at her. But that's as far as she
would allow herself to go. Touching, looking into the
eyes, it was all too dangerous for the holy. For who
knows what could happen?

All over the world, religious organizations sponsored
conferences and "dialogues," but if you ever tried to share
something tangible, like flesh, you found it impossible. To
meet, you had to abandon your own position, and here in
Israel, every group, every sect, every conclave of belief
was dug into their doctrine and surrounded by multiple
fortifications. Religious conference games were like the
nuclear round-table games, creating structures to adjust to
each other's demands, the balancing act of the world,
fighting for bones on a stage of death.

So I decided to forget it all, and just enjoy things as
they were. The house wasn't far from the sea. The air
was clear. It was quiet, and I needed a rest anyway. I
would spend a few days there before going on to Jerusalem and not expect anything from anybody.

The next day, Michelle took me on a bus to Nazareth.
We went to mass at the Basilica of the Annunciation, and
spent the day walking the stone streets, visiting Mary's
Well and other sites of pilgrimage, each claimed by their
respective sects to be the true site of the Annunciation.

We returned to Haifa and went to the cave where the
prophet Elijah is said to have hidden from Ahab, and
later walked over to the golden dome of the Bahai
Shrine. The next day, we went to Galilee and sat in
silence at the spot where Jesus had spoken to the multi-
tudes. In silence there was no disagreement. Soon, however, bus loads of American evangelists began to arrive.
They congregated on the mountain, singing songs and listening to their preachers rant on about the loaves and
fishes and the coming end of the world. There was something very discordant about the shirt-and-tie hymns as
they rose in their glory and trailed off into the air. Per-

haps it was the certainty of the resurrection, the bold timbre of hello, how do you do, are you saved? But as contraction went through me, I understood that it was my problem, my own holding, my own reaction that created any discord. I turned to Michelle. She didn't seem to be very comfortable either, despite her purity and veneration of the Christ consciousness.

On the way back we spoke of people and places. Whenever Michelle brushed against me, she would pull herself back. "Sometimes, when I'm chanting," she said, "or when I'm in church, it gets real quiet, so quiet, and everything stops." She paused for a moment and then said, "But no matter how high I go, I'll never give up Jesus." We sat quietly, now, as the bus passed Mount Tabor and rolled through the desert.

The poor Nazarene—did he know it would turn out this way, the idol of the West, the texts rewritten, the politics of hysteria?

"When you get to Jerusalem," she said, "take a walk by the walls of the Old City at night when everyone has gone. You can really feel him there."

Jerusalem

We passed through the desert. A crescent moon was shining in the royal blue sky as the night stars appeared. We sipped tea on the bus and surveyed the long sand dunes. The barren vastness of sand was dotted by occasional army fortifications. As the bus drew close to the city, my heart became anxious as I remembered those words, "O Jerusalem, Jerusalem, killing the prophets and stoning those who are sent to you."

The walls of the old city were visible in the night. They stood over the hills of Judea, revealing the great Mosque and the Dome of the Rock where the Temple of Jerusalem had once stood. How, in the name of God, could one describe the sight of that temple? "How often would I have gathered your children together as a hen gathers her brood under her wings." I felt, at that moment, as if I had achieved the fulfillment of my life. I was home in the holy city, under the wings of the Lord in the land of God, the abode of redemption.

The streets of the old city seemed so familiar, as if my soul knew them and could hear the footsteps of the Master. It was all true. It had all happened on this earth, in this city. Here, he had walked and had spoken to the people. Here was Jerusalem, where domes and crosses, crescents and stars, all came together.

The playful dance of Krishna's lila, the primordial serenity of the sitting Buddha, form and emptiness—but Jerusalem manifested differently. It was the center of the world where God had appeared in human history. But now, the house was desolate and forsaken, and who is it that would come in the name of the Lord?

By the Bab et Khalil, the Jaffa Gate, I prostrated myself and kissed the soil. Finally, I had come to Jerusalem.

A sign by the bus stop, advertising a certain hostel, had caught me eye. I knew that it was the place where I should stay. I walked back through the narrow streets to the Damascus Gate. Right outside the gate, a young woman approached me. She was wearing jeans and a blue coat. Her hair was disheveled, and her face held a look of wild determination. She told me that she was possessed by the Holy Spirit, that she had been saved by Jesus Christ and as a result had been kicked out of her hotel. Now I didn't have much money so I tried fumbling out a phrase like, "Don't worry, the Lord will provide." She lashed out in rage, "Don't you preach to me! I'm possessed by the Holy Spirit, and the Holy Spirit wants you to give me money!" She began to curse me for being American and then cursed everyone at her hotel. I walked-ed away feeling guilty and confused. Jesus said to love everybody, but if you really tried you might last ten min-

utes in this world. I should have given her something, or
have tried to put her up somewhere. I walked back to the
square but she was gone.

I was given one of eight beds which filled a small,
faded blue room at the Palm Hostel. Men or women
could sleep anywhere. Nobody seemed to care. In the
foyer, covering a chipped paint wall, were large letters
saying, "Blessed are those who come in the name of the
Lord." By the windows were stacks of Gideon Bibles being offered free of charge.

The foyer also served as a lounge. By the front desk, a
large tape deck blasted rock music of the "The Doors"
group all through the building and out into the street.
Guys sat around smoking and nodding to the music. A
girl in a low tank top and jeans did half a dance by the
desk. Others milled around by the door. In the corner
was a little stove on which tea was brewing. There were
all sorts of nationalities here. A few had picked up odd
jobs at hospitals and hostels like this one and were ready
to remain indefinitely. Thank God, there was no preach-
ing in here. Outside, hundreds of religious prophets were
just waiting to lay their hands on you. Inside the Palm,
things were laid back and relaxed.

The early Jerusalem sun filtered through the hostel
window. Everyone was asleep in the rows of beds. I tip-
toed over to the shower room. It was a small, oblong
room with faded yellow tiles and chipped walls which
rose up to a very high ceiling. The plastic shower curtain
was torn, and the structure that held it looked as if it
could collapse at any moment. As I fumbled around for
my toothbrush, I looked deeply into the mirror on the
wall for cracks, remembering that a cracked mirror is a
bad omen. At that very moment, my hand slipped and
the mirror crashed onto the tile floor breaking into
countless pieces and slivers of glass.

All morning I meditated on the meaning of the cracked
mirror. Seven years of bad luck, the tales said. I went

through one personal fear after another; illness, accidents, death, misfortune, and the rest. At the end of a Jewish wedding, a glass was always broken in remembrance of the dark side of life, the destruction of the temple of Jerusalem, reminding you that your own glass would break soon enough.

Later I went over to the Western Wall, the "Wailing Wall," and stood with the crowds listening to the outpouring of anguish and prayer. Bearded Hassidic and Orthodox Jews dressed in black suits and wide-brimmed hats swayed back and forth, dovening in one continuous moan, by the last rampart of the old temple:

> How doth the city sit solitary that was full of people!
> How is she become a widow!
> She that was great among the nations
> and princes among her provinces,
> How is she become a tributary!
> She weepeth sore in the night,
> and tears are on her cheeks:
> among all her lovers she has none to comfort her,
> all her friends have dealt treacherously with her,
> they are become as enemies.
> Judah is gone into captivity because of affliction
> and because of great servitude:
> She dwelleth among the heathen,
> She findeth no rest!
>
> The Lamentations of Jeremiah

I walked back up the stairs to what was once the Great Temple of Jerusalem, now, the Dome of the Rock. They said that you had to be Moslem to enter the mosque, so I became a Moslem and prayed, kneeling on the rug with other pilgrims. Later, I sat alone and gazed over the olive trees of Judea. The Rock, scene of the sacrifice of Isaac, the broken mirror, the broken wall, the burning roof and tower, is this how it would all end, in another religious war? "You have not kept my ways, for everyone who does evil is not good in the sight of the Lord." Would the ancient prophesies be fulfilled? Would the vengeance of the God of Israel burst forth again? Would the earth be given unto judgment? The image of the shattered glass came forth, strongly taking its place in the mind.

The vapor of hot tea and smoke filled the air in the bazaars of Old Jerusalem. Arabs with wrinkled leather skin slowly puffed on their hookas, leaning back as they sold clothing and souvenirs to the Western tourists. They didn't seem to be particularly bothered over who thought they owned the city this time. Christian pilgrims from Europe, well dressed and respectable, paraded by the old dug-up stones of the Via Dolorosa on their journey to all fourteen stations of the cross. And I thought I might remain in this place forever, and make this my eternal occupation... tea drinker. The broken glass, I would throw it off. After all, what could happen to me that hadn't happened to this city already? I took my tea and breathed the air, alive with the smells of the market place. Enjoying my breathing, I didn't want anything else.

As far as the tea drinker was concerned, the stars and their patterns of fate were of no importance. What was the need for omens, signs, and visions when one could walk the earth and witness the spectacle of life; the veiled Moslem women, the seething movement of goods, the rugs, vegetables, oils, and cheeses, the domes and vaults, the ruins of thousands of years, the costumes of all the different religious denominations, the chaos and confusion, and all that it might bring. I would let expectancies fly and make my home right here, behind a tea stall in Jerusalem. In this free fluidity the omens, signs, fortune tellers, and scriptures had to lose their grip. So the glass of Jerusalem would break again. So let it be precious, all of it.

The Citadel of David

High up by Mount Zion was the Citadel of David, a long cylindrical tower from which one could look down on all Jerusalem. Next to the tower was a stone building where the Last Supper is said to have taken place. Some fifty feet away, a long stone wall curved along the mountain. As I sat by the wall, I saw an extraordinary number of Orthodox Jews continually passing by, dressed in black coats, black hats, and long beards. I asked one of these passers-by where they were coming from and was informed that this area was the location of the Yeshiva of the Diaspora, a school for Talmudic studies which had been founded after the war.

The young man whom I had stopped soon returned with a few of his companions and asked me if I was Jewish. The whole scene was intriguing, so I nodded in the affirmative. Then I began having second thoughts. I really didn't want to go into it. Like so many other ghosts, I would have liked to believe that I had gone past it all and had attained a pure universality beyond all conditioned distinctions. I felt myself throwing up a wall of resistance. What was this curse or blessing of karma? Heredity, environment, opportunity: I knew that inside I was none of these things. But the blood of the past remained in my body, linking the centuries, the blood of minds and hands.

I sat by the wall trying to protect myself from these strange men in uniform. One of them walked over and sat down next to me by the wall. Turning and looking me in the eye, he said, "Hey, can you dig the energy around here?" Taken back in surprise, I gave him the kind of look that says, "Who the hell are you and where on earth are you coming from?" He went on with his questions. "Where you from man?"

"The United States, New York," I answered.

"Where in New York?"

"Brooklyn."

"Where in Brooklyn?"

"Bensonhurst."

"Where in Bensonhurst?"

"Bay Parkway."

"No! I'm from Bath Avenue! What high school did you go to?"

"New Utrecht."

"Utrecht! I went to Lafayette. We used to beat you guys in football every year."

The whole thing was getting outrageous, especially when I found that his cousin was an old friend of mine from the neighborhood, and that we had dozens of mutual friends from earlier times. I kept blinking my eyes in amazement at this young man in a long black coat and wide-rimmed hat running down the whole Bensonhurst scene. I could resist no longer. Dropping my psychic armor, I threw out my hands and asked him how in the world he ever wound up here and in this outfit

"I was on my way to India," he began. "That was eight years ago. I decided to pass through Israel first, you know, the 'Holy Land' and all that. It was easy then. The borders weren't closed in Iran and Afghanistan, and you could go overland. I think the train from Istanbul to New Delhi was about forty bucks. I was walking alone right down there," he pointed out over the wall, "when I got hit by this light. I mean really zapped. A bolt of light just came zooming in from nowhere and knocked me right over! I fell down and broke my leg right here," he pointed to the shin bone of his right leg. "So these guys from the Yeshiva came and put me in a *Mickveh* (a ritual Hebrew bathing tank). My leg healed instantly. Well, after that I knew I had to check this whole thing out. The Torah, it's our heritage. It's the most mystical book ever written. You should check it out too. After all, you are a Jewish soul."

"I'm what? How do you know what I am?" I reacted. He just smiled as though he had heard it all before. Still, we liked each other and we were from the same neighborhood, so we decided to spend the day together. He had been given the name Richard Greenfield at birth, but

187

was now called Reuben. He took me all over Mount Zion that morning. Stopping in the room of the "Last Supper," he told me that when the Pope had come to Israel he had offered the Rabbi millions of dollars for the building. Of course the Rabbi refused to sell.

We went up the tower of the Citadel and looked out over the desert. "You and I probably go back a long way," Reuben said. "We must have been together in the desert when Moses received the Torah." From his solemn tones, it was apparent that some of the Orthodox mystic schools believed in reincarnation. I asked him more about his philosophy and about why he had decided to stay in Israel.

"In fact," he continued in his serious voice, "I can hardly believe that I have been allowed to live here and to learn the law of creation. Were one or two things slightly different, I would probably be back in America 'doing my thing.' But, like yourself, my view was always focused on higher things, on our inheritance from our father, Abraham."

So I too was focused on "higher things." It was the usual come-on, but I was still amazed at what was happening and asked him to explain further. He pronounced the names of the founding fathers in Hebrew, "Elohai Avraham, Elohai Y'Itzchak, Elohai Ya'acov." The promised land and the covenant, to be a kingdom of priests and a holy nation with descendants more numerous than the stars. Desert wanderers in strange lands, locked in ghettos, taking refuge in the book and the way of the law. Now the twelve tribes were being called home. Now the day of prophecy had arrived. Now the Holy One, blessed be He, would again reveal Himself and smite the worshippers of Baal.

"Previously," Reuben said, "by the grace of the Almighty, I had experienced that transcendence of form, and union with that essence of Love-God-Being. Yet, I was always forced to come back down to mundane reality, to this physical world and its duality. At the time, I thought this was because of my body and sought the means to transcend it permanently. I would fast for days, practice yoga and macrobiotics. As I began learning, however, I understood that this was neither possible nor congruent with the laws of the universe. I knew that I had to make my peace with God's love and accept that the purpose of my enlightenment was to live in the physical world. The more I learned, the more I felt that I could no longer avoid this conclusion. Otherwise, at the very feet of God, I would be denying His love."

"Learning" was the terminology employed, not only for the study of the Torah and rabbinical commentaries, but for an entire way of life, a way wholly devoted to the pursuit of God and the understanding of His ways. As Abraham had argued with God over the chastisement of Sodom and Gomorrah, so the student was allowed to question every doctrine, to consider every possibility, to uncover every subtle shade and nuance of the law. The single unquestionable premise, however, was that of the first Commandment. "I am the Lord thy God who brought thee out of the land of Egypt. Thou shalt have no other God before me." This acceptance of the Eternal was the basis of all other existence.

"Through the study of the Torah," Reuben said, "I have learned to make peace with the world of matter. The traditional customs, the *Halakhah*, enable the soul to fulfill the purpose of this creation and to advance through the higher realms of light. We take meat on Sabbath, not for gratification, but because it is a *mitzvah*, the duty of a Jew to honor and thus transmute the elements. We are not here to escape the creation like the Indian ascetics, but rather to fulfill God's purpose in the creation. That is, the total integration of formlessness and the essence of love with the way of the world and its good-bad dualities. This is called *ratzoh vashav*—the dancing back and forth of the angels between the Creator, their own non-being, and the created, their identity as God's messengers. This is the highest ecstasy, and it rises on different levels of experience forever."

We remained silent for a while in the power of the

hills of Zion. Then Reuben turned around with a queer smile and said, "Do you ever wonder why everyone is always dumping on the Jews?" I made no reply. "Stick around for awhile," he assured me, "and you'll find out."

When we walked down from the tower, some black-coated young students ran over to Reuben and began to speak excitedly. Apparently one of the women in the Yeshiva had been seriously burned when an oven she was opening exploded. People were rushing to the temple to pray for her. Reuben grabbed my arm and hurried me over to the temple, which turned out to be an old classroom with a wooden arc in front. It was necessary by law to have a *minyan*, at least ten men present, in order to begin the prayers, and it looked as though I was the tenth. "Just pray," exhorted Reuben who was already swaying back and forth, dovening and crying out phrases in heartfelt tones from the Hebrew prayer book. I didn't know much Hebrew, so I mimicked his movements and asked him softly if it was all right if I chanted "Om." "Do anything you want," he said, "but don't let them hear you." In a short while, however, the temple was overflowing with people.

Later on, when things loosened up, we talked about old times. Reuben, having decided to come down to earth, was now married and had two children. Children were an important *mitzvah*. He was busy writing a book on healing according to the Torah and had collected all sorts of material on remedies, prayers, and potions from the sacred writings. He invited me to stay at his home and to sit in on classes at the Yeshiva. He also promised to arrange a meeting with the Rabbi for me. "You're a Jew. You should check it out. Everyone here has been everywhere and has done everything, and now they're just making it back home in time."

As we saw people go in and out of buildings, he told me one fantastic story after another about each of them. "This is the place," he said. "Anyone who finds his way here has been guided." He explained that after the war the Rabbi had come, driven a stake into the ground, and declared this spot to be a holy place of learning. "And that's it, we're not moving," he said. "We are going to establish a true school of sacred learning. And everyone comes here, even Bob Dylan came here a few years ago. Check it out."

The Holocaust Museum

I sat through two morning classes. The learning was intense. The literature was vast and would take lifetimes to read. The instructor was a young man in a pin-striped suit with a *yarmulke* covering his head and *tsitsith* of strands of cloth hanging down outside of his garment, which served as a reminder to fulfill the minimal amount of daily mitzvahs. He explained that the only purpose in life was to study the Torah. While it may not be too great a tragedy to have read other kinds of literature, once one has understood that a Jewish soul could attain salvation only through the Torah, not to take up its study would be tantamount to committing spiritual suicide.

In the afternoon Reuben took me over to the Holocaust Museum. Everything was on graphic display; charred bones piled one on top of the other, gaunt faces stripped bare, bars of soap made from the bodies of the dead Jews, photographs of Auschwitz and Dachau, and newspaper articles from the "night of broken glass." There was a hard fire in the eyes of the museum curator as he took me around. He and Reuben were going to make their point.

I was brought into another room filled with examples of recent eruptions of anti-Semitism. There were newspaper clippings from the American Nazi Party, articles on gatherings of the Ku Klux Klan and vigilante groups in Wisconsin who were undergoing military training and vowing to fight a "holy war" until every last Jew in America was killed. There were photos of the latest bombings and desecrations of synagogues, articles on the "International Jewish Conspiracy," the "Jews as agents of Satan," and the rest. The curator, a well dressed man in

his thirties with black hair and a strong build, was determined that I should gulp down every insane anti-Semitic
tirade from one fear-filled area after another, "He may
look like you...but he's not like you...the Jew," said one
newspaper headline. Finally we left the rooms and sat
down on a long bench in the hallway which was dark
and overcast like a dungeon. "Well," the curator said.

"Well what?"

"Well, it's up to you. Can't you see? It's happening
again, just like the last time. They said it was impossible,
that being Jewish or not was no longer significant in a
democratic society, but when they come to put you in the
ovens what will you say?" He made a pointing gesture
with his index finger. "Even if there is one drop of Jewish
blood in you, even if you deny it forever, even if all your
friends are Christians in high places, they will still call
you a Jew and take you. And do you know why this has
happened? Because the people, people like you and me,
our own people, denied their heritage. You can read it. It
is very well explained in the Bible."

The curator looked over towards Reuben who was sitting off to the side. His neck was smoldering. The room
was silent. I should have left it that way, but I just
couldn't. I knew that everything he was saying might
well be true, but for me, it was too late. I could never
live in the old way. I could belong to no country, no people, no idea. I would not cling to life, nor would I resign
to fate and death. The mystery was all, and it had to be
for all.

"You are born into a body," I answered him. "Into a
name, a designation, a religion and the rest. You may
bear it on your shoulders or wave it high in the air, but
it will die regardless, and that is for sure. Either you are
burnt in an oven, or run over by a car, or held up and
knifed in the street, or you're a victim of heart disease,
cancer, or a nuclear attack. But you will die with all
your beliefs. And when you die, what are you? Are you
your youthful exuberance, your middle-aged responsibilities, your old-age aches and pains? Are you anybody at
all?"

No one was speaking. Silence hung dead in the air. But since I had already tipped my hand, I figured I might as well go all the way. "Look, every day of our life we walk the line of the unknown. We walk in ignorance and help-lessness. We walk the line of fear, and we refuse to cross it. We cling to our idea of it all. We cling any way we can. And we are ready and willing to kill for our ideas— Auschwitz, Viet Nam, Cambodia, Uganda, Ethiopia, what's the difference? Everybody believes that 'I am right,' every ideology, every philosophy, every religion."

The curator threw up his hands in disgust and looked up at Reuben with hopeless eyes, with eyes which said, "See, this is the problem." I got ready to go and shook hands with the curator. I shook hands with the suffering and humiliation of an entire race. His handshake was hard and strong, firm and bitter, determined to never let it happen again. I flashed on the graves by the Mount of Olives and then on my own heartbreak and wretchedness, and I was frightened. I was frightened by the sagging feeling that said that we had all gone too far, that we had crushed out forgiveness forever, that men would never be reconciled with one another, that we could never be reborn.

I thought of the herdsmen in the desert. This land was a part of them. It was visible in their faces, in the way they walked through the brush and sands. Now they were being pushed aside. Now their lives and families were be-ing destroyed for an idea. Reuben seemed resigned as we walked back to the Yeshiva. He had tried his best. Now we were just friends.

Ein Kerem

I went back to the Palms, to the music and tea, to the stack of Bibles on the window ledge, to the broken mir-ror. People were still hanging out in the lobby and the tape player was still blasting the drunken music of "The Doors." I didn't want to fall into it, but I did. The music had its own power. Perhaps it was the "bhajan" of the West.

There was a woman sitting in a faded green chair by the wall just taking it all in. She wore jeans and a sweatshirt with cut sleeves. Her boots were caked over from traveling. Her name was Laura. She was a schoolteacher from Ann Arbor and had left her job to travel. We went outside together. The shops had closed and the streets were settling down. We walked out of the hostel and over to the Damascus Gate.

Laura had been in Jerusalem for some time and was very much up on its current events, although she had a sense of humor about the whole thing. "If you want to see arguments," she said, "don't worry about the Syrians or Jordanians. Just attend a meeting of the Israeli Knesset [Parliament] and listen to them yell at each other. You wouldn't believe it."

We went up a circular stairway which led to the wall. You could walk up on the ramparts and all around the old city. Now that the city had become still, you could feel its full power. The old city, the holy city, was very much present, crossed with events backwards and forwards, past and future, domes and crescents, its living history written on the old walls and battlements.

She was logical. "If there is another war," she said, "there is no hope for Israel." But it didn't rouse her blood too much. Something else had brought her here. She was wrestling with it, but it was too much for the moment. A friend of mine who had become a Zen monk had once told me that when he first met his teacher, Roshi, as he was called, had made an unusual observation about the spiritual process. "Some people take up Zen practice," he said, "to find God. Others, to get away from God."

"I realized," my friend told me, "that I couldn't get away. I had been trying to. I thought it didn't matter, but I could not get away from God. Later on, when I became frustrated with Zen practice, I had an interview with Roshi. I told him that I was leaving. Roshi asked me why, and I said that I just couldn't stand koans. I was going to become a Christian."

"Ah, but Christians have koans."

"Then I'll become a Jew."

"Ah, but Jews have koans."

"Then I'll become a Moslem."

"Ah, but Moslems also have koans."

So there was no escape. The koan, the unsolvable riddle, stared you down. You could think around it, talk around it, perform rituals around it, but it remained haunting you all the way to the foot of the cross. So we just walked around the wall and spoke on current events, leaving the haunting center alone, but knowing full well that we could never escape it.

I had told Reuben that I would meet him for dinner the next evening. His wife and children were away, and he wanted some company. But it was still early when I got up and the day was in front of me. I took a bus out to Bethlehem in the morning to visit the Church of the Nativity. The bouncing bus rocked to the rhythm of the Doors all the way through the hills of Judaea to Bethlehem.

The church, which is said to stand over the spot where Jesus was born, is shared by three different Christian denominations—Greek Orthodox, Roman Catholic, and Armenian, each with its own separate entrance. Had this been that remarkable place of that remarkable birth? Had there been three separate entryways then to mark the birth of a Jewish child?

After ducking under the low doors, which were built that way to keep out looting horsemen, you went down through the main choir to the Grotto of the Nativity. The walls were dark and covered with tapestries. Flickering candles lit the way. A small star in the eastern corner of the sacred cave marked the spot of the Advent of the Christ. I stood there, closed my eyes, and tried to feel holy.

I wandered out over to the Roman Catholic section of the church which contained the altar of the Holy Innocents and the room where Saint Jerome had sat, translat-

ing the holy Bible. The star of the Magi following the
signs of belief to a small manger and the words of the
scripture mapping the road to heaven converged here in
what had now become the land of belief itself. But outside, the townspeople sipping thick Turkish coffee were
not so concerned. They were easy going and friendly to
anyone who cared to hang out. But it was hard to just
hang out at the site of the Advent of the Savior of the
world. If you weren't holy you had to at least try to be.

I had gone to Bethlehem out of duty, but that afternoon I went to Ein Kerem out of desire. I felt deeply
drawn there. This was the "village of Juda" where Mary
had come to visit her cousin Elisabeth, and where, by the
Franciscan church of Saint John, one could find the
grotto where John the Baptist had appeared in this world.

The Baptist, that raging wielder of the prophet's staff,
ranged over the land of locusts and wild honey announcing the coming of the kingdom. The Baptist, pure power
of righteousness cut to the bone and pointing directly at
you. "You have not kept the ways of the Lord." No
irony, no sugar-coated language, just the direct burning
message. Take it or leave it, surrender unto the way of
the Lord or do it your own way and die.

I walked through the quaint village and visited the
Church of the Visitation near the house of Zacharia, but
felt more inclined to leave the village altogether and go
out into the fields. For the Baptist belonged to the
wilderness, not to the churches. To take up his staff was
to affirm—not to hope, but to believe, not to doubt, but
to go forward, to march, to change the world.

On the way out of town, I had been attracted by a
small art gallery which was one of the last buildings on
the street. I went inside, milled around, and exchanged
pleasantries with the curator. No Baptist here. As I was
leaving, a woman who had been over in the corner
turned around, and our eyes met just for a moment. Dark
eyes, Middle Eastern eyes, our eyes met for a moment of
acknowledgement, and I left.

Ein Kerem

I roamed the outskirts of the town until dusk when I remembered my appointment with Reuben. I began walking back to town to the bus station. It was getting late and the next bus would be the last one of the day. The bus sat, idling in the station, its engine slowly running. Two or three people were inside waiting. The driver had been drinking a cup of coffee and looked as if he was just about on his last gulp and ready to take off.

I boarded the bus and reached down for my money when my stomach came up through my throat. My shoulder bag was gone! Where was it? I fumbled all around me. I looked back over the bus station. I had been carrying a shoulder bag which contained my wallet, my identification papers, my money, and my notebooks. Everything I had was in that sack, and it was gone. I tightened. I panicked. The people on the bus just blankly looked at me. The driver offered to take me back into town, but I knew I hadn't left it at the hostel. Damn, it was the Doors, that's what I get for getting caught up in that music. I had forgotten who I was. I had forgotten everything. Maybe some sly thief had lifted it on the bus, or someone at the hostel who was in one of the beds near mine. No, all day long on the bus I had been bouncing up and down, rocking to the Doors. I had probably bounced the sack right out the window.

I mentally retraced every step I had made that day. Did I have the sack in Bethlehem, on the bus? I would go to the embassy. I would go to Reuben. I would call him from the bus stop, but his phone number had been in the sack with everything else. The sun was going down. "Okay," I said. "I've got to find it. I will find it." I went back through the fields. I retraced every step. I knocked on the door of the Franciscan church, which was closed. I kept banging the door frantically. A caretaker finally let me in, and I searched all through the pews. There was nothing. I went back through every street, passed the art gallery, through the town again. But now it was dark, and it was chilly, and I had nothing. All my notes, all my papers, all my money, everything was gone. I thought of calling on the saints for help, but why call for rescue from my own stupidity? Why was I always looking to be saved as if that was still the most important thing in the world? I decided to make one more round even though it was so dark that I couldn't see.

My papers, my money, that didn't matter so much to me, but my book, my best poems were in that book, my little notes and collections of addresses from my pilgrimage. I was in shock. I had let my energy get out of hand with the Doors' "Love Me Two Times" and "Light My Fire" emotionalism. I ran back into town, hoping to get a taxi, but I had no money. I made repentant vows all the way. I did affirmations. "I will find it. I will find it." The bus station was empty. There were no more buses, no more taxis, and no more people.

So I stood there in the middle of the bus station, in the middle of nowhere, and finally just dropped my arms and said, "I don't give a damn. Who needs yesterday anyway?" That's all that was in the sack, a sackful of written yesterdays. This is it. Here I am. It is now. Why not accept it? Why not just accept the verdict of the universe?

It was night. There was no place to sleep. I didn't care anymore, but I thought that just for penance, I would retrace my steps one more time. It was about an eight-mile walk back to town, and I could always ask directions and take a midnight hike or, if I was lucky, catch a ride. I walked back through the town and towards the outskirts. Just then, on a narrow street, a figure emerged from a doorway. It was the dark-eyed woman I had seen in the art gallery. She asked me if I wanted some tea.

She called herself Ila. She was dark and her eyes were deep. She was wearing a purple shawl around her shoulders and over her head. She took me in. I was sick and feverish from my search. She sat me down and got me a cup of tea. She lived here in Ein Kerem as an artist of some sort and worked at the village hospital as well. She had been here for quite some time.

The house was filled with books and Middle Eastern

paintings. There was an old sofa on a woven rug, a desk with all kinds of papers on it, and a small kitchen. Two windows looked out onto the narrow street. The room was warm, and the steam from the kettle was much welcomed.

We sat for some time. She asked if I had eaten today, and when I said no, she went in to make some dinner. Beware if a woman cooks for you! That's just the beginning. But I was dazed. I was dizzy, and my head was pounding. She came back awhile later with a plate of steamed vegetables and rice. We sat. We ate. We didn't talk much at first, but soon began exchanging stories.

It turned out that she knew all about the Yeshiva where Reuben was staying. She laughed, "Everybody knows about that place." She, herself, had originally come to Jerusalem to study with some Sufi group but had gotten out rather quickly from the "group pressure." It was then, she explained, that she had been able to explore and discover Jerusalem, and now she never wanted to leave. I had been sweating so feverishly that she had given me a blanket to cover myself, but now I felt stronger. We talked into the night about the art of living in the Holy Land. To live here, to really live and breathe the air of this presence, was to be in the constant mood of ultimate intensity, a mood which was analogous to living in the holy thirtha of Vrindavan. Only here, there was a sense of urgency, for the lila was not so much divine play as it was divine fate.

We held each other and looked into each other deeply. But the emotion was not that of two souls meeting from out of nowhere. There was meeting, and there was need, but the context was one which overshadowed any personal history or desire. Nothing else was to be done. "I'll sleep on the couch," I said. She covered me with some more blankets, and I fell out into another world.

She was already up when I opened my eyes. She smiled. We sat by the window and sipped some tea. She took out a piece of paper and gave it to me. On it was written a Sufi poem to the "beloved." She gave me some money for the bus back into town.

It was early. I wanted to catch the first bus back in and do whatever I had to do to face up to my karma. I walked down the narrow street. It was still chilly. The sun was just coming up over Judaea. I turned the corner. I hadn't even taken six more steps when my foot hit an object on the ground. It was my sack, complete with papers, wallet, notebook, and all!

The Rebbe

I returned to the Palms, called Reuben, and explained why I hadn't come to his house the night before. We planned to meet at the Yeshiva in the afternoon after which I would spend the evening at his home. I went back to the row of beds, fell down on one of them, and slept for a few hours.

Most of the students at the Yeshiva were Americans. They all kept their heads covered with yarmulkes but still clung to American remnants—blue jeans with rainbows, long hair with pony tails, and hip talk. One of them told me that he had agreed to stay at the Yeshiva and learn on the condition that he could remain on his macrobiotic diet. When he fell ill, however, he couldn't find it in himself to refuse the bowl of chicken soup that was brought in to him. After all, it was a mitzvah. He took the soup and quickly got well.

I spoke to another man who was about six-foot-six and wide too. He had rainbows all over his jeans. "This was the last thing I wanted to face," he told me, "that I'm Jewish, that I have a Jewish soul, that this is my path. A good friend of mine got me into it. For six months we stayed up, night after night, fighting it out. I resisted at every turn, but finally couldn't deny it. I can't tell you some of the trips I've had to go through here, the clothes, the food. And I really don't know if I'll stay forever, but I can say plain out to you that these people are the coolest, most gentle people I've ever hung out with. It's one big family here and it's beautiful, just beautiful."

One of the Yeshiva instructors, a Harvard man who looked like my seventh grade history teacher, Mr. Winkler, bemoaned the fact that someone like myself with my education "didn't know the first thing about his own religion." He immediately engaged me in dharma combat as he chain-smoked cigarettes and explained how the Sanskrit scriptures of India originally came from Hebrew, and that "Brahma," for example, was derived from "Abraham." I told him that it really didn't matter to me who came from anywhere anymore, but he gave me a whole bibliography and offered me a position on his research staff. *"Ha Shem* [the Holy One] is bringing his lost children home. This Yeshiva will become a great center of learning," he said.

I said that I was touched by his offer but had certain duties in the States. He had a certain humor about the whole thing and our conversation eventually turned to the academic world. I told him that in America university people were frantic because the little money they had was drying up and that Ph.D.'s were slowly taking over the taxi-driving industry. He laughed, took a drag of his cigarette, and then said, "Ha Shem's got all the money. . . . You just have to know what dance you've got to do to get it."

I spent the night at Reuben's. Getting up early, I sat in meditation and listened to him doven and recite the holy prayers in Hebrew. His wife and children were still away and the dishes were all piled up in the kitchen caked with old food. He didn't know how to cook. It was the wife's job. We got a huge breakfast together—bagels, yogurt, cheese, oranges, dates, and on and on. Reuben told me that I would be able to meet the Rebbe, the founder of the Yeshiva, in the afternoon.

The Rebbe was sitting back, leaning against a chair. His stomach and loose white shirt protruded out of his black coat. His garments were rent in mourning for his father who had just passed away. His thick lips were moist and slouched downward, and he was constantly scratching or rubbing himself. There were many wrinkles burrowed through his face, but they were good wrinkles pressed into a kindly, natural face with soft brown eyes buried deep in his large head. He looked me up and down and then made a motion that I should sit.

I said the proper words of respect for his father which Reuben had taught me before going in, and he acknowledged them. His disciples all sat around him in a half moon formation wearing similar clothes, younger versions of himself. He reminded me of a good grandpa, and he had not lost his sense of humor. "So, you have been on the India trip," he said in a familiar accent which I couldn't place. "We have many here who have been to India, did you know?" His voice meandered as he tapped his fingers on the arm of the chair in rhythm, feeling in control of the situation. "You've been on this trip and you've been on that trip, so why don't you try the Jewish trip? Eh? Just try it for awhile———after all, how old are you?"

I told him I was in my late twenties. "Oy Vey! You wander and you wander. If you go on like this do you know what will happen?" He turned and pointed his finger towards me. "You will wander around for the rest of your life and you will never settle down!" He said this as if it were a fate worse than death. I wondered what was so bad about it. Then I knew. I would not accomplish the mitzvah of having a family, which was one of the most important of the six hundred and thirteen mitzvahs a Jewish soul had to fulfill.

As the Rabbi continued, the moon of disciples nodded along in agreement. "Now why not stay here. You can get married...to a nice Jewish girl, a spiritual girl. Stay here and see. We can't keep you of course, but we'll do our best to try. If you would just give it, say, two years and see what happens." We talked on, and finally said a cordial farewell.

The Doctor

Later that evening I heard that "the Doctor" had arrived. He seemed to be an important person in the Yeshiva as I heard constant references to him. I was surprised to hear that he wanted to meet me. Maybe they were impressed because I told them that I had gone to Harvard.

He was standing outside his office in a conservative blue suit and tie. He was short but carried himself with dignity. Motioning me with his hand, he asked if I would like to take a walk. We went out by the wall which looks over Mount Zion and on to the Mount of Olives. After walking in silence for a while, he spoke. "You have come to Jerusalem. Why?" I didn't quite know what to say. I looked out over the mountains, over the temples, churches, mosques, and buildings. The long rows of graves surrounded the base of the mountain.

"I felt drawn here," I said. "There may be many reasons." I hesitated. "But I guess I came here for God." I noticed his face began to soften. I put my hands on the stone wall and leaned over. The sun was setting on the Holy City, streaking the sky.

The doctor looked up and took my wrist with his right hand. I felt a great warmth coming from his hand. "Your name is what? And your father's name? And his father's? Then your Hebrew name is Judah. And do you know what? You have a Jewish soul." I felt my reaction rising along with his words, but he cut me off. "No, 'Jewish soul' means that you are here in Jerusalem for a certain reason. You want a better world. That is why I am here. As a doctor I have practiced here for twenty years, and for the last ten years I have also been serving here and in similar places. I, too, did not choose to come from Europe. No, there is a quality in the soul that draws one here, a quality that seeks not so much for salvation as it weeps for mankind. I will tell you a story from the Talmud, a story that is rarely told."

The doctor continued slowly and clearly. His voice was deep and yet captivatingly gentle. There was a certain quality about him that reminded me of Yogi Ramsuratkumar. "Do you know the sacrifice of Isaac? Well, according to some of the great Rabbis, God never asked Abraham to kill his son. Rather, the direct rendition of the Hebrew word is, "to raise him up." That meant that God was asking Abraham to raise his son up to his own level, and Abraham could not accept that, would not stand for it, that his own son, a subordinate, would be his equal before God."

He paused by the wall and looked down on the city. "And to this day, no one else can accept it either." He pointed back towards the Yeshiva, "And that is why they cannot accept you. And *no one* accepts another as they are. We want to bend the other to our will. Abraham would rather have killed his own son than accept him on the same level as himself. So, you see, God sent the angel to stop Abraham, and Abraham saw that he was wrong and that he had to let his son go.

"The killing of the ram is the sacrifice of the evil in Abraham by the grace of Ha Shem. So I will tell you now, you go your own way, and I will go my own way. But in the soul we are one because our purpose is one." He took hold of my hand and I felt an ancient warmth, a lightness and peace descending over Jerusalem.

It is written that the Messiah will come down from the Mount of Olives. The dead shall be raised. The sick shall be healed. The nations shall be judged, and the Kingdom shall be established upon the earth. All wrongs shall be righted, unbelief shall disappear and error will be no more.

The Mount of Olives stood in the ancient sky rising over the dome of the Great Temple. A little way down was Gethsemane, now a sanctuary surrounded by trees. And under the sky were trembling hoofbeats, charging horsemen dead and fallen, and rivers of long blood. It is said that He would come down from the mountain and that the true Israel who had endured and prevailed would awaken to reclaim her purpose and take up the covenant of Light.

But there were still the wind calls, and the slow haunches of camels still trod the barren desert with no regard for human time. The leather-skinned desert men with their liquid eyes, asses laden with satchels of goods, robed women drawing water, young people looking out of windows and dreaming at the stars, the busy market place, the crumbling stones on the hill, rent garments, rivers and towers; all here, all to pass like windblown leaves, released of all.

The dim battlements of the old walled city and the temple gates were lit red in the dusk. The wind calls were coldly indifferent to all that stood in form, and from over the hillside, the graves were sunken into darkness. Waiting for the Resurrection, they had become desecrated and shorn, chipped and fallen, bare monuments of wind-carved rock.

Each monument, each tombstone left to commemorate the dead, each with its epitaph—soon the writing would wear off. The graves would be forgotten, or perhaps future generations would find some illegible script and make all sorts of suppositions about its sayings. I remembered the words of the sage who was able to teach the entire law while standing on one foot.

> If not you, who?
> If not now, when?

Epilogue:
From Wyoming to Washington

The nation's capitol was clean and airy with everything in its proper place. Disorders were not on display. Winding pieces of metal sculpture resting by fountains reflected glass buildings labeled "U.S. Treasury Department" and "Institute of American History." The Washington Monument, phallic incarnation of Shiva, rose upwards as pilgrims passed around it paying homage. A group of demonstrators had gathered on the lawn to protest the banning of school prayer, while counter demonstrators wore placards declaring it a sin to pray in public.

I had been back in the States for a number of months and was still digesting the experience of my pilgrimage. Certain things, however, had already come into focus. Above all, there was a sense of solidity and a renewed commitment to manifestation, to living strong on the earth, accepting the gift of incarnation. The quest, the image, the flight of the heart was there and would always be there. For to lose it was to lose heaven. But flight could only happen in its fullness through earth, through time, through one's given situation.

I was engaged to be married to my dearest friend and spiritual partner, Elisabet, who had shared quite a few journeys with me. We had some time off and some energy, and decided to get in a car and travel cross-country. Washington, D.C., was our final destination.

From Wyoming to Washington

The long strip of green mall between the Washington
Monument and the Capitol Building was lined with trees
and benches. We walked along with the others, having
come to pay our respects to the symbols of our land.
Along the sides of the mall, lining Constitution Avenue,
long vans opened up into shops, selling Capitol Building
T-Shirts and candy bars glossed with photos of various
sports heros.

The inside of the great domed building was covered
with pictures of fighters for freedom; frescoes of Washington crossing the Delaware, Jefferson Memorials, Hanuman lifting the mountain. After exploring the site, we sat
outside by the pool in quietness and meditated on the
dome.

The demonstrators were massing on the lawn now. A
podium had been constructed with amplifiers, micro-
phones, and a mesh of wires. Uniformed guards patrolled
the area. Other demonstrators were on the Capitol steps,
handing out cards and assorted pamphlets. I took one and
put it in my pocket as we left the area.

We had traveled across Wyoming, into the Badlands,
down through Yellowstone, and on into the deserts of
New Mexico. We had camped on mountain tops with the
U.F.O. people from Sedona. We had sat before the dancing Kachinas in Hopiland. Behind the blaring radios and
Coca Cola cans, Native American culture lay hidden,
tucked away in the desert, not to be seen by the eyes of
the white man. But one could feel a definite shift as one
moved westward. The connections with European culture
grew weaker. The land felt vibrant and younger. The
Native American power permeated this land. The astral
was filled with braves, squaws, and medicine people who
were willing to work with anyone, red, white, black, or
brown, to heal the earth and her people. One became
aware that this land had its own sacred places: rock formations of power, healing forests, and holy rivers. This
land held its own secret past, as stately and strange as the

bison and elk who looked up at us in Yellowstone. The Earth Mother and the earth cultures, hurt from violation and blatant disregard, may have turned and hidden themselves from our view, but that which turns returns again.

Fom pilgrimage it was clear that the cyclic mosaic of the sacred was all-inclusive. In some way, I had been all of these people and places, had played every role, from the most arrogant to the most oppressed. The dialectic of one side pitted against another was but the outmost layer. Within its play lay the motto *e pluribus unum*, engraved in every exchange of currency, to see the one in the many and to still *be* the many. Such a possibility was deep within this land.

We had driven thirty-five miles down a dirt road in New Mexico at night en route to Chaco Canyon. We camped out in a ravine, and in the early morning saw the ancient Anasazi dwellings within an awesome ring of cut rock formations. They spoke of the return to the circle, the clan, the community of trust, with the ritual kiva in the center. Where had these people come from? Where had they gone? What legacy was left here resonating among these ruins?

We walked through the Pueblo Bonito alone. The frontier has traditionally been the land of the lonely, and America is known as the land of the frontier. Pilgrimage itself is often a solitary walk, but from this pilgrimage, it had become clear that one must not walk alone, that the Spirit is validated through the community, and that the holy place can only become holy through the recognition and nourishment of its people.

Later, driving through Arizona and not knowing where to go, we listened inside. We once worried over where the "voices" and promptings came from, for this too has become a power game of the New Age. Cathedrals and dogmas may be gone, but the "Masters" are in, so that disciples and hangers-on can walk new roads to slavery, flocking around the one who is "in contact with the ascended Masters, the Archangels," and so on. But be-

yond exotic power, beyond the uncommitted hope for a
new mode of experience, the open heart *does* lead into
another dimension of intuition, moving into the unknown
with a shared sense of openness.

Pilgrimage can and does open this sense, especially
when one sees the holy rivers and mountains, the towers
and temples to be, not of "God" alone, but rooted in the
people and their experience with the holy place. It is the
awesome contact of the Spirit with her people that creates
the holy monument, the sign that can be read through
the heart and through the people of the heart.

We wound up in a health food store in Cottonwood,
Arizona. It was eight o'clock in the morning, but the
door was open. When we walked in, a bearded worker in
striped overalls asked if we wanted some tea and showed
us a place to sit down. The gates of dharma shall open,
the holy places shall be unveiled, and the people of the
Spirit shall recognize one another. We asked if they knew
Shawn and Aurelia. They gave us a phone number. We
were unannounced but not unexpected. For here, there is
always place for another, and phone calls are made in the
mind.

It was blazing hot that afternoon. We went out into
the desert, Elisabet and myself, Shawn and Aurelia, and
their two-year-old son Raphael. Shawn drove through the
heat and dust and stopped his truck in what seemed to be
the middle of the desert. We got out and followed him
through the hot sands and down the rocky brush of a ravine until we came to a long, hidden river. Only those
who knew the land knew about this river. It could not be
found on maps. We stayed on the banks for the rest of
the afternoon, lying in the sun and wallowing like water
buffalo in the cool waters, swimming upstream and making eye contact with the fish, just being with the deep
rhythm and clear-green waters. Raphael didn't need to
whine or cry to become the center of attention. For the
center was the river, and its depth was magnetic, holding
everyone together.

Aurelia was very close to the Indian vibration, moc-

casins and beads, a sense of community, and a deep love
for the land. She was preparing the house for an evening
sweat ceremony. Brother John would lead. Some people
were coming from down the road, and others from miles
away. Shawn built houses for a living and had built this
series of homes on the Arizona plain. Those living here
were of like mind. There was no sect or religion and no
fences between the separate dwellings. There was no rhet-
oric of alienation or smug spiritual superiority. There
were just folks living together with other folks, trying to
live in vision. Raphael could walk outside and wander.
Wherever he went, someone would keep an eye on him.
Here, you didn't have to keep out your neighbor, and you
didn't have to make an appointment to come over for
dinner.

People began arriving as the sun went down. Aurelia
had put the wood on and began heating the sweat lodge.
Later the rocks, over which the water was to be poured,
were heated. A massage table was set up in the back for
anyone who wanted to be worked on.

Everyone gathered and sat in the lodge together.
Steam, mixed with the smoke of burning sage, rose from
the center, its vapor circulating through with the wooden
walls. Men and women sat side by side, unheard of in
traditional Navajo sweats. In any traditional culture, for
that matter, men and women together in a sweat lodge
was unthinkable, as unthinkable as the Hindu chants that
began to resound inside the Native American sweat lodge,
as the cross on the living room wall made of holy cedar
wood.

But this was America, mongrel, naive, and outrageous.
And in its naivete came a possibility of freedom and an
extraordinary creativity. Being unbound by any specified
tradition or culture, the entire world—the Indian, the African, the Native American, the Celtic, the Pagan, and
the Christian—merge into a new form, a form which is
at once unique, and yet sprung from the nurturance of
the world past. The pilgrimage through form does not

end with an atavistic fixation or with a denial of form itself. It continues with its process of creation, opening to
the new, the challenge of joyful responsibility.

The smoke rose with the chanting. The drone mixed
with the vapor and scent of wood. Brother John stood in
the center and spoke. "O Grandfathers, create in me a
humble heart." He poured the water out as the steam
rose from the rocks. Sweat poured. The air thickened.
The heat rose. It was hard to breathe but we all hung
together. "O Grandfathers, may I not be a prisoner in my
own mind, in my own ways." The heat was intense. Im-
purities poured from the body and mind. Some sang.
Some sat in silence. The air was still. As in a Quaker
meeting, whoever was inspired spoke. "O Great Spirit,
teach us to honor this land."

The water table all over this area had been shrinking
for years now. The situation was serious, but most ig-
nored it, doing loads of wash, leaving taps running. Here
the water was used sparingly. Every drop was precious
and every person knew it. "O Great Spirit," a sister
called out, "please keep us in your current. May we not
walk away. May we not lose your divine current."

In the silent current, in the smoke and vapor, hands
clasped other hands. Envy and fear rose up through the
smoke and heat and evaporated. Here no one was lord
over another, but the Lord was present. Here no one
feared humiliation, but the wonder of nature made one
humble. Here there was nothing to become, nothing to
make, but by gathering together, a stage was being set
for the coming of a new way of living, one which would
be a work of art, a craft to be carefully mastered with
patience replacing time, and with a sense of shared
destination.

Whenever you felt ready, you could leave the lodge.
Outside, the air was clear, and billions of stars shone
overhead. Some jumped into the cold pool nearby and returned to the heat of the lodge. The air was filled with
presence. When I finally walked out and sank into the

pool, I felt so clear and cleansed, so relaxed and clean. There was no worry, not a shred of grasping, of posing, of becoming, just the sounds of night crickets and slow desert winds, the breathing of the universe, your own breathing, the transparent air, the luminous moon, the great expanse.

We returned from Arizona driving up through the South and stayed with Elisabet's parents for a week, honoring our most immediate roots, the heart of our own mosaic, the family. We then finally traveled up to Washington, D.C. After so many places and so many lives, our paths converged here. The pilgrimage was the sustenance. The challenge, the possibility, was here.

When we left the reflecting pool by the Capitol Building, we went over to the Theodore Roosevelt Bird Sanctuary island to relax and be with nature. The sky was clear and silent. The birds, trees, and clouds had their own movement. Low flying planes came in every few minutes, drowning out the bird calls over the Potomac. We sat amidst the droning planes. Directly across the river, the Watergate apartment complex was visible. Towards the right, one could see the traffic running by the Washington Memorial Parkway. I took the card from my pocket which the demonstrator had handed me and read:

> I will lift up mine eyes to the hills
> From whence cometh my help.
> My help cometh from the Lord.

To blend the old and the new, the one and the many, the individual and the community, this may be a path. Here, in America, living itself is pilgrimage. The frontier is spiritual. The frontier is corporeal; Indian canyons and data-bases, desert fathers and cybernetic computers, Spanish sanctuaries and NASA rockets, African healing powers and silicon micro chips, Asian art forms and laser screen projections, Hassidic chants and multimedia videos, Krishna, Christ, and Allah right on the streets of the walkaday world. Here living shall be my pilgrimage, and here, whether in winged sandals or basketball sneakers, I can walk the road to God.

About the Author

Rick Jarow left his studies at Harvard University in 1970 and traveled throughout Europe, spending time in various spiritual communities, ashrams, and Benedictine and Trappist monasteries. He then settled in India for some years where he pursued various spiritual disciplines, including Raja and Bhakti Yoga. He became a Research Fellow at Braj Academy in Vrindavan, where he studied the Vaishnavaite Bhakti movement as well as Indian languages and literatures.

Since his return to the West in 1978, he has sought to combine the ancient disciplines of the East with the knowledge and insight of Western traditions on both a theoretical and practical level, being particularly concerned with the creation of new cultural vehicles for spiritual expression. Toward this end he has taught classes and led workshops in meditation and the healing arts. He has also completed a higher degree in Indic Studies and Comparative Literature at Columbia University, and has served as the Dean of Continuing Studies of the New Seminary in New York City, an interfaith seminary dedicated to the training of spiritual counselors and interfaith ministers.

For the past four years the author has been a fellow in the South Asian Department at Columbia University in Sanskrit and Comparative Literature and is currently an instructor of Oriental Humanities. He has written on such diverse topics as "Eastern Saints and Western Culture Heroes," "Parallel Death Experiences in Eastern and Western Traditions," "Vrindavan—The Holy City of Krishna," and "A New-Age Fable," and various New Age poems. He has been awarded a Fulbright scholarship and expects to return to India to continue his research.

Quest publishes books on Healing, Health and Diet, Occultism and Mysticism, Philosophy, Transpersonal Psychology, Reincarnation, Religion, The Theosophical Philosophy, Yoga and Meditation. Two popular titles from the above categories include:

The Silent Encounter *Edited by Virginia Hanson*
A book about your mystical nature and how and why you could suddenly become aware of life's unity.

Whispers From the Other Shore *By Ravi Ravindra*
How religions help and hinder us in the search for our center of being.

Available from:
The Theosophical Publishing House
306 West Geneva Road, Wheaton, Illinois 60187